P9-DWP-908

BODY SHOPPING

BODY SHOPPING

THE ECONOMY FUELLED BY FLESH AND BLOOD

Donna Dickenson

ONEWORLD
OXFORD

A Oneworld Book

Published by Oneworld Publications 2008

A CIP record for this title is available
from the British Library

ISBN 978–1–85168–591–2

Typeset by Jayvee, Trivandrum, India
Cover design by James Nunn
Printed and bound by
the Maple-Vail Book Manufacturing Group,
Braintree, MA, USA

Oneworld Publications
185 Banbury Road
Oxford OX2 7AR
England
www.oneworld-publications.com

Contents

Preface

Is your body just a consumer good, like any other? Can your genes and tissue be processed, sold and turned to make a profit? They most certainly can, and any number of interested parties have their eyes on them, in ways that will probably never have occurred to you. Part of the job of this book is to alert you to what they might be.

Unlike some other exposés of organ sales and the tissue trade, this book doesn't conclude that the way in which the body is becoming an object of commercialisation —'body shopping'—is inevitable. It can be resisted, it is already being resisted in many parts of the world and it should be resisted elsewhere. Nor does the book lump together all the instances of body shopping as equally monstrous or momentous. Some are more important to resist than others, although these are not necessarily the ones that have attracted the most media attention. Everyday, less dramatic instances—like umbilical cord blood banking—probably affect more people than florid abuses.

By taking a historical long view—for instance, comparing the agricultural enclosures of common land in early modern times and the patenting of the human genome—and a global broad view—with examples from Tonga, France, the US, the UK, Spain, Cyprus, Eastern Europe, China and elsewhere—this book enables the reader to make those careful distinctions. It's easy to feel lost in a big buzzing confusion about all the new technologies, when some fresh headline about new advances in stem cell research, or whatever, hits the papers every week. But even if the technologies are new, the moral problems they raise are familiar enough to be within our grasp. Looking at international and historical comparisons helps us get a purchase on them.

What this book also enables the reader to do is to recognise the ingenuity of the 'body shopping' industry in all its unexpected forms, from private umbilical cord blood banks, to the trafficking of women for their eggs in southern European clinics, to the better-known but still shocking cases, such as the international trade in the organs of executed Chinese prisoners. The landscape is hard to map because the shorelines and continents seem to be drifting all the time, with every new development in biotechnology. This guidebook is as up-to-date as it's possible to be in such a shape-shifting field.

In many cases, I've been fortunate to have had an unusual degree of access to the scientists and doctors pioneering these developments. For example, Chapter Seven, on cosmetic surgery and face and hand transplants, draws on my experience on the UK hospital clinical ethics committee that was asked to approve the first human hand transplant, and on my experience as a co-presenter at a seminar at the *École de Médecine* in Paris with the French surgeon who performed the most recent face transplant. Chapter Four, on the Hwang Woo-Suk fraud in the stem cell technologies, includes little-known accounts by the Korean feminist activists who were instrumental in bringing the scandal to light. Colleagues from several European Commission projects have also provided me with rarely publicised material, for example about the extent of commercial egg sales in Spain.

I have done my best to get the science and medicine right, based on my long experience of teaching and writing in the field of medical ethics and law, as well as working with physicians on hospital ethics committees, in professional bodies and in UK medical Royal Colleges. At the same time, I've been careful to make the legal and ethical analysis as painstaking as possible. So although it should already be clear that I don't think body shopping is a good thing, I do present arguments in its favour: for example, the claim that commercial biotechnology will present us all with manifold benefits if we let well enough alone. I also try to ground those arguments in the ideas of philosophers, from Plato to Locke to Marx.

My training as a philosopher and lawyer tells me to play devil's advocate to my own beliefs, and that's what I try to do. But I do also take a position. It's not the commonly held one in the field of bioethics, where most academics, at least in the English-speaking countries, favour organ sales, for example. So I've had to argue all the harder for it.

This book is also different in that it isn't just about the United States. The US is still (thankfully) atypical, although probably the furthest developed in 'body shopping' (a strange way of being the leader of the free world, but there it is). Perhaps that's because of its health care system. Since US residents with no health insurance regularly have to make decisions about how to set the value of their health against their cash, Americans already think in terms of their body parts as 'worth' dollars and cents. For example, an uninsured man who lost two fingers in an accident was offered a choice of reattaching his ring finger for $12,000, or his middle finger for $60,000. He could only afford one, and that had to be the cheaper option.[1]

The notion that a particular body part is worth *x* dollars clearly underpins a Faustian health care bargain like that, even though organs and tissues (apart from eggs and sperm) can't legally be bought and sold in the United States. In European socialised medicine or insurance-based systems, that routine equating of body parts with cash doesn't have to occur to anything like the same extent.

And that matters, because the standard US response, on both sides of the political spectrum, to new risks—like those to women who donate eggs for stem cell research—is to assume that if we have paid people enough, they have accepted the risks voluntarily and can be assumed to have accepted the consequences too. But that argument just won't wash. In other areas, like health and safety at work, anyone but the most rabid free-marketeer would accept that the fact of being paid isn't the end of the debate: there is still a rightful role for regulation to protect the vulnerable. We need to look at other models of regulation of body shopping, and there are some promising ones in the academic literature, like the notion of a biobank as a charitable trust run for the benefit of the patients and the public. Part of the role of this book is to bring some of these models to a wider audience.

Body shopping is indeed alarming in its pace and scale, but excessive scaremongering has the adverse psychological effect of making us too frightened to take action against its worst abuses. As the French say, *du calme*, and as they also say, *bon courage*. The fight isn't over yet. Indeed, it's hardly begun.

Donna Dickenson
Paris and Oxford
June 2007

Acknowledgements

In writing this book I've had a tremendous amount of help, provided altruistically by colleagues and friends all over the world. Although the gift relationship is frequently abused in biotechnological research, I've benefited from the genuine article.

Proceeding chapter by chapter, I should begin by thanking Prof. Itziar Alkorta Idiakez of the University of the Basque Country for extensive information on the highly commercialised Spanish IVF clinics discussed in Chapter One. She and other European scholars with whom I worked on the European Commission-funded Network for European Women's Rights (NEWR) have helped me to break out of the Anglophone bioethics ghetto. My thanks also to Prof. Hille Haker of the University of Frankfurt, another colleague from that project, for inviting me to give a lecture in June 2007, the first on the final draft of this book. I am also grateful to my very old friend Pauline Rummens for providing me with material about the scandal concerning Cooke's bones.

Professor Lori Andrews, who has been involved in most of the major US cases on commodification of the body from *Moore* onwards, is another NEWR acquaintance who deserves thanks, especially in relation to Chapter Two and Chapter Six. Lori and her colleague Professor Julie Burger very kindly sent me the hot-off-the-judicial-press decision of the Appeal Court in the Catalona case, just as the final edition of the book was due. For their very timely assistance, and their original analysis of the question of continuing rights for patients in their donated tissue, I would like to express my gratitude, respect and admiration.

Chapter Three, on commercial cord blood banking, includes

clinical information to which I would never have had access but for the involvement and support of Dr Susan Bewley, then chair of the Ethics Committee of the Royal College of Obstetricians and Gynaecologists. I respect a great many clinicians for their commitment to serious ethical debate, but perhaps Susan most of all. My thanks should also go to the librarians at the Royal College, to my colleagues on the Ethics Committee, and to my former student Saskia Tromp for first alerting me to the issue of cord blood during our supervisions.

Sarah Sexton, of the activist organisation The Corner House, provided me with extensive material, international contacts and a sounding board for my ideas about egg donation in stem cell research, all of which were absolutely crucial to Chapter Four. She also kindly put me in touch with Young-Gyung Paik of Korean WomenLink, who also deserves thanks here for filling me in on the story that didn't emerge in the newspapers, about how Hwang's fraud really came to light—in large part through the activities of Korean WomenLink. Diane Beeson, Professor Emerita of California State University, East Bay, was very helpful in providing further relevant material. My thanks also to the organisers at the ESRC Innogen Centre, University of Edinburgh, for bringing us together, on a panel on egg donation held in September 2006, and for opening this important topic to a wider public.

Chapter Five includes material on popular resistance to genetic databanking in Tonga, which I could never have learned about without the remarkable opportunity graciously afforded me by Nga Pae o te Maramatanga, the New Zealand National Institute for Research Excellence in Maori Development and Advancement. My deepest thanks to Mera Penehira, Sharon Hawke and Paul Reynolds of Nga Pae o te Maramatanga for their good company and excellent organisational skills in putting together the conference where I met Lopeti Senituli, leader of the Tongan resistance. I am very grateful to Lopeti for presenting me with a copy of his paper on Tonga at this conference and for his helpful answers to my questions.

For Chapters Five and Seven, I was also very fortunate to hear rarely publicised details of the second French face transplant from the lead surgeon himself, Professor Laurent Lantieri. My deepest thanks to my good friend Dr Simone Bateman for arranging this seminar in May 2007, and for inviting me to speak as well. During my earlier stay in 2004 in Paris at the Columbia University Institute for Scholars at

Reid Hall, where I gleaned the basis for the material on the French resistance to 'body shopping' in Chapter Five, I was given a great deal of support by Danielle Haase-Dubosc and Mihaela Bacou. Former and current members of the French CCNE (*Comité Consultatif National d'Éthique*) were equally generous with their time: among them, Nicole Questiaux, Simone Bateman and Anne Fagot-Largeault, to whom I am also grateful for her invitation to present a seminar on my work in progress at the *Collège de France*. Jean-Paul Amann, her deputy, was enormously helpful in setting up and chairing the session. At the CCNE library near the Invalides, I was warmly welcomed by staff and benefited from their excellent collection of bioethics literature, as well as from the specialised search facilities which they graciously make available to foreign scholars. Jennifer Merchant, professor at the *Université de Paris II Panthéon-Assas*, gave me a very great number of valuable 'leads' into the French bioethics and biolaw literature, which is still too little known outside France.

I would also like to express my deepest thanks to the jury and organisers of the International Spinoza Lens Award, of which I was the 2006 winner—particularly Marli Huijer and Rene Foqué, for the way in which they have helped me to see the need to bring my academic work on commodification of the body to a wider audience and for arranging media interviews in which I began to see how I could do so in a book like this.

Among many other colleagues whose comments have helped me to refine my ideas, I would particularly like to thank Catherine Waldby, Susan Dodds, Susan Sherwin, Françoise Baylis, Carolyn McLeod, Cynthia Daniels, Alex Capron, Richard Huxtable, Ruud ter Meulen, Tony Hope, Mike Parker, Geertrui van Overwalle, Debora Diniz, Liliana Acero, Cecile Fabre, Melinda Cooper, Anjali Widge, Catriona MacKenzie, Jane Kaye, Carole Pateman, Ingrid Schneider and Roger Brownsword. Thanks also to Denis Campbell of the *Observer* for alerting me in our interview to aspects of the HFEA decision in February 2007, as well as to Jenni Murray of Radio Four's 'Woman's Hour' and her team for inviting me to discuss this issue and others on several occasions.

You wouldn't think someone who had written twenty books would need to have her hand held, but Marsha Filion, my editor at Oneworld, has gladly done that for me, in the most insightful way. My agent, Peter Tallack of Conville and Walsh, London, also deserves some serious thanks for his support at all stages of the project, from the very earliest days. Additionally, I am very grateful to my website designer, Lisa Taylor of CSS Web Design, for helping me to refine the key points of this book into the 'frequently asked questions' on my website, www.donnadickenson.net, and for her patience in putting up with frequent changes when new biotechnological developments emerged.

It just remains for me to do the conventional but heartfelt thing and thank my family: my husband Chris Britton and my son and daughter Anders and Pip Lustgarten. Somehow they managed to remain unperturbed by my subject material, even at the Christmas dinner table, and to keep me convinced that I really did need to keep at it.

I am dedicating this book to the memory of the producer Alison Tucker, with whom I worked on many issues of medical ethics, first during her period at the British Broadcasting Corporation and my time at the Open University, and then during her successful recasting of her career in terms of freelance production. We produced many scripts, radio and television programmes, videos and CD-ROM scenarios together, over a fifteen-year span. Alison died at the age of fifty-one in March 2007. I could have had no better colleague or firmer friend.

1

Body shopping at both ends of life: babies and bones for sale

Trade in human tissue, like any other consumer commodity, now stretches from the time before birth to the treatment of the body after death. This chapter shows that trade—part of what I call 'body shopping', the way in which the body has become a commodity—in practice at both ends of life, arguably robbing birth and death of whatever sacred quality they still possess in a secular society. It also asks how we can understand the way in which the body has become an object—a thing—and why some commentators actually think there's nothing wrong with that.

'A GLOBAL MARKET IN BABY-MAKING'

On the recreation room bulletin board in a Spanish university, a poster urges 'Help them! Give life!' The target audience is cash-strapped female students, who are being asked to sell their eggs to a profit-making fertility clinic for 1,000 euros. A little emotional appeal to altruism—'Give life!'—helps the advertising campaign, perhaps, but the eggs aren't really a gift: these women sell their eggs. In doing so, they take their part in an expanding global market in baby-making, as do the couples who buy the eggs through IVF clinics. 'In these cases, and thousands like them', as the American commentator Debora Spar writes in her book *The Baby Business*, 'the parents aren't motivated by commercial instincts, and they hardly see themselves as "shopping" for their offspring. Yet they are still intimately involved with both a market operation and a political calculation.'[1]

Spain numbers a total of 165 private fertility clinics offering *in vitro* fertilisation (IVF)—more than any other European country.[2] Many offer astonishingly good results, far better than the dismal rates of success that IVF often produces. The secret of their success is that they have stopped employing surplus eggs from other IVF users (in the process euphemistically known as egg 'sharing') because these women are 'too old'. By definition, this hard-nosed reasoning goes, any woman attending an IVF clinic has a fertility problem. Instead, the for-profit Spanish clinics target young women at the peak of their fertility, such as students, and pay them for their ova.

So do the many private US clinics that pay for eggs. Advertisements like the one on the Spanish university bulletin board regularly circulate in US college newspapers, offering egg 'donors' amounts up to $50,000,[3] from an average of about $4,500.[4] 'Desirability' of genetic traits primarily determines the price: blonde, tall, athletic and musical donors command the higher rates, at considerable risk to themselves. One report documented the taking of seventy eggs at one time from a 'donor' who nearly died in the process.[5]

The US market for fertility treatment operates on a gargantuan scale. Americans paid well over $37 million for 'donor' eggs in 2002 alone.[6] Monies paid to egg sellers, however, were dwarfed by revenues to drug companies for fertility drugs (over $1.3 billion) and to IVF clinics (just over $1 billion). The Center for Egg Donation in Los Angeles, the first commercial egg 'brokerage' service, opened for business in 1991, followed rapidly by larger brokers like the Center for Surrogate Parenting, the Genetics and IVF Institute and the Repository for Germinal Choice. A full-fledged market has now emerged, with a differentiated pricing structure following geographical trends: highest in New York, lowest in the mid-west. The Center for Egg Donation boasts an online database, from which clients can shop for 'donors', viewing photos of the egg supplier and her children, reading about her hobbies and even checking her SAT (college entrance exam) scores.[7] 'Boutique retailers', such as the Californian company A Perfect Match, place very specific advertisements in Ivy League college newspapers, such as their 1999 offer of $50,000 for eggs from a seller who was at least 5'10" tall, had a combined SAT score of 1,400 points and possessed a blemish-free medical record. No doubt the price would be higher now.

In January 2007, it was announced that a for-profit 'human embryo bank' centre was even offering one-stop shopping, eliminating the need to select eggs and sperm in separate transactions. A Texas company, the Abraham Center of Life LLC of San Antonio, became the first firm to provide batches of embryos from which customers could choose their preferred model off the peg. Selecting only sperm donors with a higher degree and egg donors in their twenties with at least a college education, the Abraham Center nevertheless denied any taint of eugenics: 'We're just trying to help people have babies,' said director Jennalee Ryan. She, together with some commentators, differentiated between producing babies to order, custom-made, and offering customers a choice off the shelf; the first smacks of eugenics, creating a 'master race', but there's nothing wrong with the second, they argued, because choice is a good thing.

While some bioethicists condemned the new embryo bank as blatant baby shopping, others remarked that it was just a logical extension of choosing an egg donor or a sperm donor by their genetic characteristics or educational level. John Robertson, of the University of Texas at Austin, shrugged off criticisms: 'If you step back a little bit, you realize that people are already choosing egg and sperm donors in separate transactions. Combining them doesn't present any new major ethical problems.'[8] But *is* it fair enough to shop for eggs and sperm as if they were consumer goods and, logically, to produce embryos and babies made to order? The president of the American Society for Reproductive Medicine, Steven Ory, remarked of the Abraham Center: 'We find this very troubling. This is essentially making embryos a commodity and using technology to breed them, if you will, for certain traits.'[9]

In the world of for-profit fertility, the Abraham Center is just offering a tailored product, in a market where branding is everything. Another niche market, gay men, is met by a new Los Angeles group, The Fertility Institutes. According to Dr Jeffrey Steinberg, director of The Fertility Institutes, 'We are the only program for gay men that has psychological, legal, medical, surrogates, donors and patients all taken care of in one place.'[10] Gay couples can select the sex of their offspring (or, strictly speaking, one partner's offspring), with about 65 per cent requesting boys. The majority of initial clients weren't US citizens but rather men from Britain, China, Canada, Italy, Brazil and South Africa.

Private American fertility centres mimic commercial companies in their advertising, treating their clients like customers and babies like any other consumer good, available on demand. At the Advanced Fertility Center in Chicago, for example, customers are offered a money-back guarantee: no baby, no payment. To produce such a medically improbable result, when the average pregnancy success rate in US clinics is 27 per cent,[11] this clinic must be treating patients and 'donors' with dangerously high levels of hormone stimulation. No matter: the Center's Affordable Payment Plan 'can make your fertility care less expensive than a second car'.[12]

But that's what we might expect of a free-market economy with an ethos of personal choice. If Spain, still a profoundly Catholic country, displays a similar kind of free market in human eggs, it suggests that 'body shopping' is rapidly evolving into a global phenomenon. Human tissue, including human eggs, is increasingly just another object of commerce, and that phenomenon occurs around the world. Just as 'in today's global market, a healthy human egg from a young white European woman is more valuable than gold', [13] so other forms of human tissue and genetic material are the focus of a new 'Gold Rush', whose Klondike is the human body.

The world-wide scale of egg selling, as an example of globalised 'body shopping', applies to both sellers and buyers. The US market in eggs began as a mainly internal enterprise, but by 2003 approximately one-third of customers at the US Center for Egg Donation came from abroad, often through the global Internet.[14] Conversely, American women are among the customers of Southern and Eastern European for-profit treatment centres. Most external demand for Spanish private clinics' eggs, however, comes from Germany and Italy, where egg donation is forbidden by law, with an estimated three thousand German women obtaining Spanish eggs every year. But even the comparatively liberal United Kingdom, which allows egg donation but forbids payment beyond a maximum of £250 for 'expenses', has begun sending couples in search of eggs to Spain—not deliberately, but as a side effect of the 2005 policy abolishing anonymity for egg and sperm donors. Many British couples now travel to Spain to get round that requirement.

Nor are the sellers of eggs to Spanish clinics necessarily Spanish themselves. Immigrant women, mainly from Eastern European countries, provide an important alternative source of donors to female students. Now that the Iron Curtain has been drawn aside, Eastern European women are 'free' to sell their eggs anywhere in Europe. And so they do, particularly in Cyprus and Spain, Southern Mediterranean countries that act as a point of transit between East and West.

At the Petra Health Clinic in Larnaca, Cyprus—an offshoot of the Reproductive Genetics Institute in Chicago—women recruited through the clinic's branch in Ukraine are paid $500 to fly in and 'donate' eggs. The clinic's resident Russian director, Galina Ivanovna, claimed in a 2006 interview with the *Observer* that these women were being given an all-expenses-covered holiday, not paid for their eggs, although her account was a little confused. 'We put them up in flats and give them a free holiday but now, it seems, they feel they can pay for their own. If you wish,' she told an undercover reporter, 'you can pay them too.'[15] In return, the 'client' would be allowed to choose from a range of donors according to preferences in height, weight, hair and eye colour, education level and occupation. 'Do you want a baby who looks like you?' Ivanovna asked: personalised baby marketing.

Although many of the egg 'donors'—better termed sellers—are unemployed or working in menial jobs, female engineers and other highly educated women can also be drawn by the sum offered: paltry in Western eyes, but enough to live on for six months in Russia or Ukraine. Larissa Kovoritsa, a liaison nurse linking Russian donors to a fertility clinic in Nicosia, Cyprus, told the *Observer* reporters that some women lived primarily from selling their eggs. 'For them it's like giving blood; you give and then you forget,' said Tatjana, a twenty-eight-year-old tour representative who had considered selling her eggs but shrank back from the thought that 'there might be a piece of me, some little Tatjana out there in the world'. Not everyone is equally squeamish. 'They just give their eggs and get the money. It's a pure transaction.'[16] Coming from Poland, Lithuania, Latvia, Estonia and other newly capitalist states of Eastern Europe, these women, Tatjana claims, sell more than just their eggs. 'They work the cabarets, they'll sleep with men, they'll sell their eggs, and then they go back again.'

Blood, of course, is infinitely replenishable in a healthy individual. By contrast, it is generally agreed that a baby girl is born with all the egg follicles she will ever possess, so that each batch of eggs taken is gone for good. What the long-term effects are on these women's fertility and chance of premature menopause is anyone's guess. The phenomenon of egg selling is still comparatively recent, and the sellers mostly in their twenties, so it will take at least fifteen years for the risks of premature menopause to be known.

We already know that other risks, to do with the intense hormone stimulation to which these women are subjected, can in some cases be fatal. What's more, whereas IVF clinics in Western Europe and the United States are moving towards policies of minimal hormone stimulation, the Eastern European and Mediterranean egg-selling clinics routinely extract three or four times the quantity of eggs that would be taken in a well-run clinic. Women are actually given a productivity bonus if they produce high numbers of ova. In one Kiev clinic, for example, women are offered a basic fee of only $300, but given a bonus of $200 if they produce as many as forty eggs. Doses of follicle-stimulating hormone at more than twice the recommended maximum level are routinely used to produce these bumper crops.[17] But the human female is programmed by nature to produce only one or at most two eggs per cycle.

So given that egg sellers in Cyprus are usually paid about one-fiftieth of what the buyers pay the clinic, this form of body shopping—shading over into 'baby shopping'—looks thoroughly immoral, exploitative and shocking. So says the former Chair of the UK Human Fertilisation and Embryology Authority, Suzi Leather, who has condemned what she calls 'a global market in baby-making ... a profoundly exploitative and unethical trade'.[18] Yet other commentators see nothing wrong with this and other instances of body shopping—the way in which organs, eggs, sperm and other forms of human tissue are bought and sold on global markets like commodities. In fact, many regard body shopping as a positive force for good.

EXPLOITATION, JUSTICE AND FREEDOM OF CHOICE

There are two common responses to the way in which human tissue is becoming a commodity just like any other. The first approach, more commonly heard on the left of the political spectrum, runs something like this:

> What do you expect? We live in a consumer society, where money is the measure of all things. Bodies and parts of bodies are no different. Yes, of course, it's dreadful, but only the terminally naïve are shocked by it. You'll never be able to regulate it, either. There's too much at stake for the big biotechnology firms, and they can make life very uncomfortable for any government stupid enough to try.

The second viewpoint shares with the first an assumption that you can't buck the market—but regards that as a good thing. That view, more common to the political right, is usually couched something along these lines:

> Yes, we do live in a free-market economy, which will bring us great things if we just let well enough alone. Biotechnology is one of those great things, and it shouldn't be regulated by government. Any attempt to do so will subvert the progress of science. If selling eggs or other forms of tissue improves the fertility of women who have to undergo IVF, and also provides the sellers with an income, then that has to be a good thing for both parties. And if it occurs on a global scale, so much the better: more backward countries can be brought into the realm of the market, and their people will also benefit. It's paternalistic and condescending to interfere with anyone's free choice to buy or sell body parts.

You might be surprised to find the second view predominating among the academic community in bioethics (the study of moral and legal issues arising from the new biotechnologies) but so it does. This book is an exception to the rule. But oddly enough, most academic bioethicists who subscribe to the free-market view regard themselves as the valiant mavericks, even though they are increasingly in the majority.[19] Some of them include bioethicists from poorer countries, who might be expected to be sensitive to global injustice. Yet one Iranian commentator, for example, claims that it would be a new form of colonial exploitation to deny Iranians the right to sell their kidneys, either domestically or on a world market.[20] Apart from writers from particular religious traditions—such as Jewish commentators who interpret their *halacha* (law) as making it wrong to take advantage of another's poverty by buying his organs[21]—it seems to be harder and harder to find anyone willing to condemn the globalised trade in human tissue. At the same time, as the next section will show, the abuses of that trade are becoming more and more flagrant.

But whereas it might look obviously unjust for a poor woman to sell her eggs, at a knock-down price, to a rich couple, some commentators actually argue that justice *demands* that we allow organ sales. Cecile Fabre, Professor of Law at the University of Edinburgh, thinks that if we feel those who lack material resources should be given equal shares with the wealthy, then we ought to allow those who lack full health to have access to the organs they need to make them well. (Fabre doesn't deal explicitly with infertility, which isn't necessarily the same as illness—not life-threatening illness, anyway.) Government regulations should allow organs to be redistributed along set lines, but in addition, some types of markets in organs should be permitted, she thinks.

Can we simply equate human organs with objects of property-holding like savings, stocks and shares? We would have to do so, in order to accept Fabre's parallel between redistributing wealth and redistributing health. In fact, the law doesn't recognise any property in the body: you can't actually own your organs, tissues or eggs, in a legal sense. After all, we simply *are* our bodies; we aren't embodied in our savings accounts or shareholdings. I can't exist apart from my body, although I can exist apart from my savings account.

So Fabre's argument looks problematic from the start. We can't be obliged to share something in which we don't have a property, and we don't have a legal property in our bodies. But even if we did accept her argument about fair shares in health, we still might jib at her further claim that the sick ought to be able to *buy* healthy organs if government agencies or health services can't provide enough of them outside the cash nexus.

In fact, claims Fabre, the current system is unfair to donors who derive no profit from their organs, when for-profit clinics and private organ brokers do make money from them. The altruism of tissue donors is already being exploited. Additionally, she says, it's the organ *recipient* who is at risk of being exploited, if the pressing choice is an organ transplant or death. Those who need urgent transplants would pay any price they could possibly afford. For this reason, Fabre prefers a regulated system of government compensation for organ donors, rather than for-profit firms of the kind that predominate in Spanish and American egg provision.[22] But if that system can't meet all the demand, she is willing to countenance private sales. Furthermore, she thinks that even those who aren't in greatest ill-health have the right to pay for tissue, and that sellers have the corresponding right to provide them with organs for cash. 'One also has the right to sell one's organs to those who do not *need*, but rather want, treatment requiring body parts.'[23]

Yet it's hard to see how any such government compensation systems could operate across international borders. Body shopping is now a global phenomenon, although international regulation lags behind. It seems perverse to think that Third World kidney sellers have a duty to make First World kidney patients better, in the largely unregulated global tissue trade.[24] But that would be the implication of Fabre's view, if extended globally.

Even within Europe, where the EC Tissue Directive now provides some regulation, trafficking in human organs, just like the market in eggs, tends to prey on the poorer ex-Soviet countries outside the European Union. A report published in 2003 documented an extensive brokerage network involving organs from Russia, Ukraine, Georgia, Bulgaria and Romania (the latter two then outside the EC). Organised crime was involved in this as in other forms of trafficking, such as trafficking for sex. Although fourteen out of fifteen EC states had made it illegal for their nationals to buy and sell organs from each other, only one out of fifteen (Germany) prohibited its citizens from travelling to other countries to buy organs; the practice dubbed 'transplant tourism'.[25] The rest of the European countries have one law for their own people and another for foreigners.

Whether or not it means that the buyer rather than the seller of organs is the vulnerable party, Fabre is perceptively right to remark on the way in which recipients of tissue will be willing to pay almost any price. An eBay auction of a healthy human kidney attracted worldwide bids of up to $5.75 billion before being revealed as a fraud. In body shopping the usual rules of market transactions and elasticity don't seem to apply, and that also applies to 'baby shopping', particularly the market in eggs and sperm. As Debora Spar writes, 'For the baby market does not operate like other markets do. There are differential prices that make little sense; scale economies that don't bring lower costs; and customers who will literally pay whatever they can.'[26]

It's fair enough to bear in mind that the buyers can also be exploited in the body shopping business, but are individual egg sellers the real exploiters? Or is it the middleman? Private infertility clinics in the United States typically charge between $6,000 to $14,000 for each cycle of IVF treatment, and most women purchase more than one cycle. The mark-up for sperm is even greater: a gross mark-up, in the US, on average, of 2,000 per cent. Men receive an average of $75 per specimen, containing between three and six vials of sperm, whereas the sperm banks sell each vial for somewhere between $250 and $400.[27]

If there is exploitation in egg and sperm sales, are such price differentials the source of it? Or is the imbalance in income between the typical organ or egg seller and the recipient the real problem? Is there a problem at all, if sellers voluntarily accept the prices they're offered? After all, isn't that the essence of a free market? Aren't we just being hypocritical when we try to distinguish between selling bodily tissue at the going rate and selling any other good or service in the market?

Many commentators would say so. If both parties to the transaction in eggs or organs are happy with it, who says there's anything wrong going on? Isn't it actually liberating for both sides? Women who buy eggs can extend their fertile period, cheating the biological clock. Women who sell eggs are just earning their living by a more extreme form of what most people have to do: sell the labour of their bodies. There's nothing inherently exploitative about that, many argue: it's just a fact of life. But is that overly simplistic? We need to look more closely into some definitions of exploitation, a term that sometimes seems to generate more heat than light.

The German philosopher and economist Karl Marx wrote extensively about the meaning of exploitation, providing insights that turn out to be relevant to the way in which human tissue has become a product in twenty-first-century biotechnology—although he himself was writing about the factory goods of nineteenth-century industry. Bearing that limitation in mind, it's worth making a short excursion into what Marx had to say, before returning to another, even more extreme form of body-shopping at the end of this chapter.

Marx distinguished first between attributing 'use value' to something, *objectifying* it, and, additionally, making it an object of exchange, *commodifying* it. Objectification is just the process by which something external to ourselves is made to satisfy our wants and needs, which isn't inherently objectionable (to coin a pun). Part of what seems shocking in body shopping, however, is that our bodies aren't usually conceived of as external to ourselves. If I sell my eggs or my kidney, then I am objectifying those parts of my body and, additionally, commodifying them, turning them into objects of trade. Again, that may or may not necessarily be wrong, but until the advent of modern biotechnology, it was largely unknown. With the exception of the pan-European trade in saints' relics during the Middle Ages, there wasn't generally money to be made from human tissue, certainly not on today's global scale.

Modern biotechnology also muddies the clear distinction between things external to our bodily selves and those intrinsic to us. Mechanical ventilators or pacemakers are incorporated from outside into our bodies, while parts of our bodies such as tissue samples or DNA swabs may be separated from us for other uses. The notion of 'external' has become deeply problematic in modern bioethics. With that development come difficulties that Marx didn't have to confront: what can be rightfully separated into an object with use and/or exchange value, and what can't?

Although some analysts contend that Marx viewed commodification as wrong in itself, others assert that neither objectification nor commodification is intrinsically malign in Marx or anywhere else.[28] What is wrong is making a saleable object of something that should be treated as having value in itself, irrespective of what use might be made of it. Because people have value in themselves, parts of people, you might think, would be particularly problematic. If it's wrong to make people into objects or things—as slavery does—and if the body is the person, then is it wrong to trade in bodies and their parts?

So the first question is whether bodies and their parts are the sorts of things that have value in themselves, beyond the realm of commerce. If human tissue can't be turned into a commodity without harming people's worth as persons, then *any* form of tissue sale is in a sense exploitative, *whatever price is offered for it.* It's demeaning to human dignity, treating the person like a thing. In that case, there's no injustice done when the donor isn't recompensed for her eggs or organs, because human tissue isn't the sort of object on which a financial value can be set. The injustice lies in paying for the tissue, not in who gets the payment or how much the payment is.

However, it also seems unjust when biotechnology companies, for-profit egg brokers and private IVF clinics charge recipients of that tissue a price above the minimum that reflects their labour in processing the tissue. That implies a different definition of exploitation: taking away the rightful reward that should belong to the person who does most of the work. Exploitation would then have to do with the *disparity* between the amount of labour or value put into the organ, egg or sperm by the person selling it and the final price paid by the buyer for the 'finished' product. The mark-up for sperm might be less unfair than the amount offered for eggs, even though the percentage of profit is higher, given the fairly minimal amount of effort and risk involved in giving sperm, compared to egg donation.

The seventeenth-century English philosopher John Locke thought property rights flowed from 'mixing' our labour with the raw materials of the production process. Marx built on this idea, interpreting labour as adding crucial value to raw materials. In essence, this 'labour theory of value' underpins his definition of exploitation. If the person performing the most labour receives the least return from the final product, then elements of exploitation have crept in.

A great deal of academic ink has been spilt in the United States about what would be a 'fair price' for eggs and sperm.[29] As we've already seen, however, there's almost infinite price elasticity for the 'finished product' (if you can call a baby a product). The 'use' value of a baby to the contracting couple in a surrogacy transaction, or to an infertile woman buying eggs through an IVF clinic, borders on the infinite. Where the line is drawn doesn't depend on willingness to pay but on ability to pay. So there are difficulties in applying Marx's analysis, because the 'baby business', and 'body shopping' in general, don't entirely fit the model of factory goods with which Marx was primarily dealing. In another sense, though, the potential for exploitation is even clearer when the 'product' can be sold for such a vastly inflated price but the 'worker'—the egg supplier, for example—receives a fixed, very minimal fee.

Marx rightly reminds us to be alert to the typical power and wealth differential between buyer and seller in body shopping. That generally operates against the seller rather than the buyer, as Fabre would have it. When the trade in human tissue is globalised and largely unregulated, as it is in human eggs, those power and wealth differentials are increased, as when poor women from Eastern Europe supply their eggs to wealthy couples from Western Europe.

It's also important to note that people can be exploited even if they sell their labour voluntarily at the price they're offered. That's why almost all modern Western governments have minimum wage legislation, or health and safety at work standards. Yes, you voluntarily choose your job—up to a point, given that we all have to live—but that's not the end of the matter. There is still a rightful realm, accepted by employers and employees alike, for government regulation. 'Free choice' is not a knock-down argument.

Clinics like Petra at least pay their egg suppliers something, even if it's a paltry sum. But there have also been a number of well-documented thefts of eggs and other forms of human tissue, sometimes at the most august institutions. Even those who favour legalising the sale of eggs and other forms of human tissue should be troubled by these cases. Does legalising a market in tissue make illegal activities like theft and black markets more or less likely? Those who favour legalisation optimistically reckon it would drive out illegal activities, but in the United States, where buying eggs is legal, illicit scandals still occur.

The University of California at Irvine has been accused of multiple thefts of eggs and embryos at its fertility clinic, dating as far back as the 1980s. Layne and Rosalinda Elison are among those claiming they were robbed. In 1987 Rosalinda was twenty-six years old, with two children, when she went to UCI fertility doctors to reverse a tubal ligation (sterilisation). Her doctors, Ricardo Asch and Jose Balmaceda, waited about eighteen months before performing the minor surgery. During that time, Rosalinda Elison said, Asch and Balmaceda told her that her eggs weren't viable and pumped her full of fertility drugs. 'I was used as a lab experiment, a lab rat,' she said. Fertility clinic records show that seven of her eggs were removed without her knowledge and given to another woman, who subsequently gave birth to twins.[30] Rosalinda didn't find out her eggs had been stolen until 2002. Along with twenty-eight other couples who allege their embryos or eggs were stolen, she has initiated a lawsuit for fraud against the university, which brazenly argues that too much time has elapsed since the alleged offences for the case to be valid.

Those who favour organ sales, even a regulated trade, might well want to distance themselves from the extreme abuses now rife in 'body shopping'. And so they should, because some of those abuses are very extreme indeed—such as the case of Alastair Cooke's bones. That example shows that no one, no matter how well-off or famous, is exempt from the abuses which this book is about. In Victorian times, it was the poor whose bodies were particularly at risk from 'resurrection men', better known as grave-robbers. In an ironic form of democracy, now we're all equally vulnerable. And whereas in the case of egg sales it was poor women from Eastern Europe who were most at risk, the Cooke case shows that 'body shopping' makes no distinctions of gender or geography.

THE UNLOVELY BONES

> That torso that you're living in right now is just flesh and bones. To me, it's a product.[31]

In December 2005, it was revealed that a body parts ring, including surgeons and undertakers, had removed the thigh bones from the corpse of the well-known broadcaster Alastair Cooke and sold them for $7,000 to a company supplying dental implants. During his working life, Cooke enjoyed huge popularity on both sides of the Atlantic for his long-standing BBC programme on US politics and culture, *Letter from America.* In death, his bones themselves became a letter from America: a warning of what happens when free markets in human tissue slide out of control.

Like the global market in women's eggs, the illicit trade in human bones is world-wide. Illegally harvested bone from the United States has turned up in dental implants and orthopaedic transplants in the United Kingdom and elsewhere. Although the UK had already had its own tissue scandal in 2001, with the Redfern report on the retention of dead children's tissue without their parents' consent by a pathologist at the Alder Hey hospital in Liverpool, no world-wide commercial trade was involved. Van Veltzen, the pathologist involved, had kept the tissue for his own idiosyncratic use.[32]

By early 2007, however, a considerable number of patients in the UK had undergone surgical procedures using bone from the US criminal ring involved in the Cooke case. In September 2006, twenty-five UK hospitals were warned about the recall of some eighty-two suspect bone products by Swindon-based Plus Orthopaedics, a company connected with the dubious US supply chain centred on the New Jersey firm Biomedical Tissue Services. More than a hundred criminal counts of forgery, fraud and grand larceny have since been lodged against the firm's director, Dr Michael Mastromarino, and an embalmer connected with him, Joseph Nicelli.[33]

In most of those cases, the UK hospitals came clean and notified the patients concerned, but three major London teaching hospitals refused to inform their patients that they might now be contaminated through what was meant to be a healing procedure.[34] Alastair Cooke died at the age of ninety-four from lung cancer, which had spread to his bones. As Cooke's daughter, Susan Kittredge, said: 'That people in need of healing should have received his body parts, considering his age and the fact that he was ill when he died, is as appalling to the family as is that his remains were violated.'[35] Later she wrote: 'Imagine for just a second being told by your doctor—as thousands of patients have been—that in retrospect they aren't exactly sure where the tissue they put in you came from. How could you run away from yourself fast enough?'[36]

Those UK hospitals which did notify their patients offered them screening to rule out infection with hepatitis, HIV or syphilis. In the case of Cooke's bones, however, there was an additional risk of contamination, because the stolen thigh bones would have been affected by cancer. Not only were his thigh bones pilfered as his body lay in its casket in a Manhattan funeral home; his records were also falsified by the New Jersey firm, with his age wrongly certified as eighty-five and his cause of death recorded as cardiac arrest, not cancer.

Nor was Cooke the only victim. Over a thousand other bodies were targeted by the same New Jersey ring, which is alleged to have been operating for at least five years in an extensive conspiracy including undertakers, surgeons and biomedical companies. Cooke's thigh bones were allegedly sold for more than $7,000, despite their cancerous condition, but other parts of the body are also in demand: tendons, ligaments and possibly even skin. In a macabre way, Cooke was fortunate: other corpses were much more extensively ransacked. The body of an eighty-two-year-old woman, Esfir Perelmutter, was exhumed to reveal that most of her bones below the waist were missing, replaced with plastic plumbing tape. Like Cooke, Perelmutter died of cancer, but her medical records were falsified to read that she had succumbed, at sixty-five, to a heart attack.

It would be comforting to think that the Mastromarino ring was a particularly grisly aberration, but the American journalist Annie Cheney has discovered that it is only one small cog in a nation-wide 'bone machine'. Before the criminal investigation into its activities, Mastromarino's firm, Biomedical Tissue Services, was part of a national network of tissue banks supplying Regeneration Technologies Inc. (RTI), a profit-making Florida processing firm that earned $75 million in 2003 alone. Traded as a legitimate firm on the New York Stock Exchange, RTI takes a 'proactive' approach designed to overcome awkward seasonal fluctuations in its 'raw material', human corpses. By courting funeral directors—known as 'crystal partners'—with the promise of amounts up to $7,000 per body, and through buying up non-profit-making tissue banks, the firm has successfully broadened its 'supplier base' to include some three hundred funeral homes across the United States. Expanding overseas with distribution agreements in Germany, Austria, Switzerland, South Korea, Greece, Italy, Spain and Portugal, RTI has turned its operations into a globalised business.[37]

Because, in many US states, no one keeps tabs on the way corpses are treated before burial or cremation, there is room for widespread abuse. Bodies intended for cremation, like Cooke's, are particularly vulnerable to mutilation, because there is no evidence afterwards, only ashes. After his death in 1955 and before his cremation, Albert Einstein's body was ransacked for his brain by the pathologist who conducted his autopsy, Dr Thomas Stoltz Harvey. In any case, a mere 10 per cent of US states inspect crematoria, and roughly half have no laws governing cremation at all.[38] The US Food and Drug Administration allegedly turns a blind eye to infringements of the law, treating data about a tissue bank's operations as proprietary commercial information. (In fact the FDA had inspected Mastromarino's company and apparently knew perfectly well that he obtained body parts from funeral homes.) Tissue banks themselves hide under the ironic cover of the donor's dignity, when pressed to reveal whether their sources of supply are fully documented and completely consensual. 'Discussing such details could give donor families the wrong impression, tissue bankers say—it could make families feel as if their loved ones were nothing more than commodities.'[39] Precisely.

Most people probably assume that body-snatching was successfully relegated to the realm of horror films by legislation against the abuses of the eighteenth and nineteenth centuries, but in the twenty-first century new sources of demand have created renewed sources of supply. The uses of human tissue have expanded to include bone dust paste in periodontal surgery, transplant of dissected heart valves, cadaver skin grafts for burn victims, and beauty treatments such as facial injections. Aborted foetuses from the Ukraine are routinely used in 'rejuvenating' treatments given to wealthy Russian women.[40]

'Suppliers' of bodies and body parts include morgues, medical schools, tissue banks, for-profit firms, funeral homes and crematoria. It may seem surprising that medical schools figure on this list, but so they do: in March 2004 the director of UCLA medical school was arrested for illegally selling donated body parts given by people who had thought they were altruistically leaving their bodies to science.[41] Scandals involving sale of bodies from willed-donor programmes have also surfaced at the University of Pennsylvania, Tulane, the State University of New York at Syracuse and a number of other American medical schools who have all worked with body-brokers. Between 1998 and 2004, Louisiana State University medical school, for example, earned nearly a quarter of a million dollars by selling donated cadavers. 'LSU has, in essence, become a corpse wholesaler,'[42] even though it is illegal for a non-profit institution such as a university to generate revenue. These interstate sales violate Louisiana state law, which makes it an imprisonable crime to transfer any body out of the state, but the Louisiana attorney general has declined to investigate.

'Buyers' are found among major teaching hospitals, medical associations, doctors and researchers. 'The demand for bodies and parts surpasses the supply, which keeps the prices of human flesh and bones very high. Each corpse that travels through the system can generate anywhere from $10,000 to $100,000, depending on how it is used.'[43] Table 1 breaks down the prices commonly paid in the United States per 'unit' of body tissue.

Table 1 Prices for Body Parts (taken from Annie Cheney, *Body Brokers: Inside America's Underground Trade in Human Remains*, 2006)

Head	$550–$900
Head without brain	$500–$900
Brain	$500–$600
Shoulder (each)	$375–$650
Torso	$1200–$3000
Forearm (each)	$350–$850
Wrist (each)	$350–$850
Hand (each)	$350–$850
Leg (each)	$700–$1000
Knee (each)	$450–$650
Foot (each)	$200–$400
Cervical spine	$835–$1825
Eviscerated torso	$1100–$1290
Torso to toe	$3650–$4050
Pelvis to toe	$2100–$2900
Temporal bones	$370–$550
Miscellaneous organs (each)	$280–$500
Whole cadaver	$4000–$5000

Tissue processing had already become big business by the 1980s, with the founding of two highly profitable companies, CryoLife and Osteotech. The less legitimate side of the business, illicit supply by funeral homes, began about the same time, when David Sconce, director of the Lamb Funeral Home in Pasadena, California, was found to have removed teeth, eyeballs and hearts from bodies intended for cremation and sold the tissue to a biological supply company.[44] 'Inspired' by Sconce's example, funeral director Michael Brown later set up a willed-body programme at his own Californian crematorium, dubbing the new operation Bio-Tech Anatomical and offering clients free cremation in exchange for body donation. Bio-Tech Anatomical then sold their body parts, without their advance consent or that of their families. Donor confidentiality meant that buyers never saw consent forms, so no questions were asked. Orders placed through brokers are even more anonymous: clients may not have any idea where the body parts originated when a middleman is involved.

Brown made over $400,000 from sales of body parts before being charged in October 2003 with sixty-six counts of mutilation of human remains and embezzlement.[45] Although he pled guilty to all charges, he exhibited no remorse. 'One way or another someone makes money off of the dead,' he said. 'Funeral homes, they're all for profit. When you drive by a funeral home and you see those signs that say that stuff about dignity and care? There's no dignity in death.'[46] Despite his own indictment, Brown doubted that there could be consistently effective regulation of the trade in human remains. 'It would be an arduous task to try and regulate it ... It's not going to happen ... Not in a capitalistic society ... There's too much money to be made.'[47]

True enough, after Brown closed up shop, his clients had little trouble in finding new suppliers, such as the Arizona firm ScienceCare Anatomical. Its director, James E. ('Jimmy') Rogers, was in fact 'inspired' by Brown, just as Brown had been by Sconce. When Rogers and Brown met, Brown's firm was firing on all cylinders, and Rogers was quick to emulate its success. 'Jimmy was like a rocket off the launching pad,' laughed Brown afterwards. 'He took it and went with it. I don't know whether it was the money or his own entrepreneurial spirit that got him to do it. But you know, the entrepreneurial spirit can't be tamed.'[48]

Opening in June 2000, ScienceCare operated an aggressive marketing campaign for donations in newspapers, senior citizens' conventions, nursing homes and hospices, with a Yellow Pages listing under 'cremation'. Offering its 'suppliers' free transportation for the body, free filing of death certificates and a free cremation, ScienceCare has quickly expanded its 'buyer' list to include major surgical equipment companies. It now has a branch in Denver and a spin-off company, operated by a former employee, BioGift in Oregon. (Abhorrent as it may seem to tout for business among those who can't afford a proper funeral, even the American Medical Association has proposed offering relatives a $10,000 tax credit or a funeral expense supplement if they will donate the body for transplantation.[49])

Although the sale of human organs and dead bodies is outlawed in the United States by the Uniform Anatomical Gift Act of 1987, the tissue 'industry' takes advantage of a legal loophole permitting 'harvesters' to charge unspecified 'processing' fees. Rogers has been careful to operate within that loophole: in a letter to Brown, he noted that 'This is another good reason to charge procurement and processing fees, etc., as opposed to fees for a specific tissue.'[50] By inflating the amount they spend on labour, transportation and storage of organs, body 'brokers' can make a tidy profit.[51]

Isn't it just a matter of closing that loophole? Those who favour legalising the sale of organs might well think so. They would probably say that it's unfair to tax them with abuses as gross as those in the Cooke case, or indeed in other shocking instances, such as the well-documented sale of thousands of executed Chinese prisoners' organs for the global transplant trade.[52] Some commentators claim that it's actually the prohibition on organ sales—which, unlike sales of eggs and sperm, are banned in the US—that drives desperate recipients into an undercover trade and 'transplant tourism'.[53] Indeed, some of them argue, it's only by legalising the sale of all forms of human tissue, eggs and organs that we can hope to bring the 'industry' out into the clear light of regulation and to eliminate black markets in tissue.[54] Illicit traders, in this sanguine view, will be caught and prosecuted once legitimate traders have an interest in seeing stricter oversight of the entire tissue industry.

But this argument runs counter to common sense and historical evidence. The simplest way to prevent abuses in the tissue trade is to outlaw the for-profit tissue business altogether, and to use the full vigour and rigour of the law in prosecuting offenders. Markets in eggs and sperm have been permitted in the US for over twenty years, but the abuses are actually getting worse all the time. We have plenty of evidence to show that so-called legitimate traders aren't policing their more doubtful *confrères*. Instead, what once seemed dubious—as designer embryo shopping does now and as the sale of eggs itself did twenty years ago—just becomes more mainstream.

It's unnecessarily pessimistic to say nothing can be done about body shopping. Outside the US, action is already being taken at both national and international level, for example, in the 2004 Human Tissue Act in the UK and the European Union Tissue Directive, both of which came into effect in 2006. There are loopholes in the UK legislation—eggs and sperm aren't covered, for example—and of course there will still be some abuses. There will always be people who break any law. But imagine if someone were to argue that because there will always be murders, we should relax the laws on murder, or abolish them altogether. That would be a very dubious logic, but it's exactly the same kind of argument used by those who favour legalising tissue sales, because black markets in tissue will go on otherwise. It's a weak-willed way of appeasing lawlessness, rather than trying to regulate it.

Once the body is viewed as a full-fledged commodity, we will lose our sensitivity to abuses like many of the cases in this chapter. Then it will become much harder to draw the line, as proponents of regulated body shopping want to do, between rightful and wrongful kinds of trade in bodies. Why shouldn't dead bodies then be viewed as one of the rightful objects? Or embryos ranged in a bank like dresses on a clothes rack? Drawing fine lines, like the one between 'custom-made' and 'ready-made' embryos, will more readily become the order of the day, once we admit that body tissues can legitimately become commodities. And some of those lines will be very fine indeed.

Traditionally, as we've seen, the law took the view that bodies and body parts were not the kinds of things that could be owned, still less made the objects of profit. But at present, as Fabre and others rightfully argue, the law allows some people to make a profit from human tissue—everyone except the person who donated it. Those who favour legalising tissue sales want to rectify that anomaly by allowing everyone, including the original 'sources' of the tissue, to buy and sell bodies and body parts within the law. The more obvious way out of the contradiction, however, is to enforce the older prohibition fairly across the board.

But why does the law take the view that human tissue isn't a thing that can be owned? And how consistent is that position? In the next chapter we'll explore the roots of the traditional view and its application in some modern examples, beginning with the case of John Moore, who protested that 'My doctors are claiming that my humanity, my genetic essence, is their invention and their property. They view me as a mine from which to extract biological material. I was harvested.'[55]

2

What makes you think you own your body?

In 1976, John Moore, then thirty-one years old, developed a rare cancer called hairy-cell leukaemia. Over the next fourteen years, he was to be transformed from patient to patent: US patent number 4,438,032, for the 'Mo' cell line, licensed for $15 million to the biotechnology company Genetics Institute Inc. and the drug firm Sandoz Pharmaceuticals.

Moore was neither the first patient used in this way, nor likely to be the last. Along with his cell line, now renamed 'RLC', tissue bank catalogues list the HeLa line—derived from Henrietta Lacks, who had died in 1951 of an usually virulent cervical cancer. A biopsy sample from her tumour, taken without her consent, was subsequently turned into a remarkably productive cell line for cancer research. On the very day she died, the lead researcher involved, George Gey, appeared on national television holding a vial of cells he simply called 'HeLa', announcing that from them 'it is possible that … we will be able to learn a way by which cancer can be completely wiped out'.[1]

The promise of scientific progress first materialised from the HeLa cells in a different form, however: they made it possible for researchers to grow the polio virus in sufficiently large quantities to enable the vaccine to be developed. The cells were so powerful that they could reproduce an entire generation every twenty-four hours. Gey freely shared his resources with colleagues elsewhere in the United States and abroad, who used them to search for leukaemia cures, to study the causes of cancer and to explore the effects of drugs. But Henrietta Lacks's widower and five children knew nothing about any of this until 1975, when her daughter-in-law Barbara Lacks happened to meet someone at a dinner party who remarked, 'Your name sounds so familiar. I've been working with some cells in my lab; they're from a woman called Henrietta Lacks. Are you related?'

By then, Henrietta Lacks had been dead for twenty-four years; her family was deeply shocked to find that her cells were still alive. No matter how much benefit the HeLa line had done—and there is no denying that it had produced benefits, becoming the standard medium for work by molecular scientists—the family has never reconciled itself to being kept in ignorance by researchers. Interviewed forty years later, Lacks's husband expressed many of the same concerns that would be raised by the Moore case: 'As far as them selling my wife's cells without my knowledge and making a profit—I do not like that at all. They are exploiting both of us.'[2]

The difference is that Moore filed a court action to recover profits from what he claimed was his property in his own body. On that ground he lost, although he succeeded in establishing that his doctors had breached their professional duty and denied him the opportunity to give genuinely informed consent. But the courts were unwilling to contravene the traditional legal doctrine that we do not in fact own our bodies, as most people probably think they do.

THE CASE OF JOHN MOORE

Although *Moore* was an American case, it was based on the common-law principle that we cannot be said to own any tissue that has been taken from our bodies. That tenet applies to the law in all the English-speaking countries. Similarly, under the civil law systems of France, Belgium and the Netherlands, for example, tissue removed during a procedure is considered to be abandoned.[3] In both cases, contracts in bodily tissue and materials are difficult or impossible to enforce, although for different reasons. In both systems, patients have no further property rights in their tissue once informed consent to its extraction or donation has been given. So although it is legally binding only in California, Moore's case transcends its immediate time and place.

Moore had consulted a specialist at UCLA Medical School in California, Dr David Golde, who performed a battery of tests—extracting samples that included sperm, blood and bone marrow aspirate.[4] Golde and his research associate, Dr Shirley Quan, rapidly realised that Moore's blood-products and other tissues augured profits. Specifically, Moore's tissues produced unusually copious quantities of T-lymphocytes (white blood cells controlling the production of lymphokines, proteins which regulate the immune system). If researchers could isolate the genetic 'blueprints' from Moore's tissue, they could then manufacture lymphokines for both research and therapeutic purposes. And like the bottomless porridge pot in the fairy tale, the 'Mo' cell line could produce unlimited quantities.

In due course, the UCLA doctors also removed Moore's spleen, which was clinically necessary. (It weighed twenty-two pounds, over twenty times the normal weight.) The clinicians had providently made prior arrangements to have sections of the organ taken to a research unit, without Moore's knowledge. After the spleen was excised, Moore was asked to revisit Golde's clinic in Los Angeles periodically, flying down from his home in Seattle to donate a variety of other samples which he was told were essential to his follow-up treatment. When Moore asked Golde whether these samples could be taken by a doctor in Seattle, Golde offered to pay for his plane tickets and a hotel stay in the posh Beverly Wilshire.

Initially Moore complied with Golde's requests, but seven years after his surgery, he began to suspect major mischief. During his clinic visits, he had been asked to sign a so-called 'consent' form that went well beyond the well-established requirement of informed consent to the operation itself, which Moore had willingly given. Now he was also being asked to sign a statement that had nothing to do with the operation and everything to do with commercial gain:

> I (do, do not) voluntarily grant to the University of California all rights I, or my heirs, may have in any cell line or any other potential product which might be developed from the blood and/or bone marrow obtained from me.

Moore began by circling 'do,' but his reasons cast doubt on whether his 'consent' could be termed voluntary, as informed consent must be. 'You don't want to rock the boat,' he remarked in a later interview.

'You think maybe this guy will cut you off, and you're going to die or something.'[5]

On his next visit, Moore was given an identical form to sign, but this time he circled 'do not'. When he arrived back at his hotel, he received a phone call from Golde, asking whether he had ticked the wrong box by mistake and requesting him to come in to sign 'correctly'. Moore didn't confront Golde directly, but neither did he go back to the clinic. On his return to Seattle, Moore received a letter with a fresh copy of the form, marked at the crucial spot with a sticker saying 'Circle "I do"', followed a few days later by a letter from Golde urging him to sign. It was then that Moore decided to take legal advice. His lawyer, Jonathan Zackey, discovered that in 1981, two years previously, Golde had *already* filed for a patent on the 'Mo' cell line, along with certain proteins produced by the cells. Golde and Quan were listed as inventors of the cell line, with their employers, the Regents of the University of California, named as assignees.

In 1984, Moore instituted a lawsuit against Golde, Quan, the biotechnology firm Genetics Institute Inc., the drug company Sandoz Pharmaceuticals and the Regents of the University of California. In addition to claiming lack of informed consent and breach of 'fiduciary duty' (the obligation of doctors to act in their patients' interest rather than their own) Moore alleged that he had been the victim of a legal wrong called 'conversion', which entails illicit interference with someone's rightful property—in this case, Moore's excised tissue.

You might say that the profit-making product was no longer Moore's property, because the finished cell line created by the researchers differed from the 'raw material' of Moore's tissue.[6] But the point of the conversion action was to try to establish that laws about wrongful use of property were still relevant. After all, Moore had explicitly refused to sign a form saying that he relinquished his rights to Golde and his collaborators. The fact that Golde had sent Moore that form implicitly proved that Golde's side recognised that Moore had property rights interests in the cell line, about which they felt uneasy.

Moore's primary objective was to vindicate his own dignity; his secondary aim was to stake a claim in the profits generated by the patent and the cell line. Those goals, however, were only attainable if the court recognised that there could be such a thing as property rights in the body. This was a historic claim. The nearest precedent was probably *Doodeward v. Spence*,[7] an Australian case in which a freak show proprietor successfully reclaimed the body of a two-headed foetus, which had been confiscated by the police. In that gruesome instance, however, the tissue donor—the pregnant woman—was not involved, unlike the *Moore* action. In the event, the final court judgment in *Moore* denied that there were any convincing precedents supporting the argument that tissue removed from the body can be called a form of property.[8] Nor was the court about to set one now. Moore failed to establish ownership rights in his tissue.

So what makes you think you own your body?—as Golde's side argued, with powerful effect, against Moore. Most people are surprised and somewhat shocked at the thought that Moore might not have owned his body. Legal doctrines, under both civil and common law systems, have left us with something of a vacuum. But in fact it *is* generally recognised that we do not own our bodies in law: they are not the subject of property rights in any conventional sense.[9]

We have a right to give or withhold consent to an operation, certainly, but that's a different matter from controlling the use of any tissue removed during the procedure. Once tissue is separated from the living body, our common law generally assumes either that it has been abandoned by its original 'owner', or that it is and was always *res nullius* (no one's thing), an object belonging to no one when removed.[10] Under previous circumstances, the tissue would have been presumed to have been removed because it was diseased, and so of no further value to the person from whom it was extracted. In that sense, it might conceivably be called 'waste', as Golde's lawyers argued and as the California Supreme Court accepted. But with the advent of modern biotechnology, the case is radically altered.

As the first chapter of this book demonstrated, people's tissue now has tremendous potential 'biovalue': at least, some forms of tissue from some people, including T-cells from John Moore. The California Court of Appeal, which ruled in Moore's favour but was overridden by the majority in the state's Supreme Court, remarked, perceptively, that the 'extraordinary lengths' to which Golde had gone in luring Moore back to give specimen after specimen demonstrated that he certainly didn't regard Moore's tissue as junk. The same could be said about the pressure Golde put on Moore to sign the release form: using the 'waste' analogy was hypocritical in those circumstances. At the time the case reached the Supreme Court, the cell line was valued at three billion dollars. Some junk!

The distinction between the patient's definite right to withhold consent to an operation and his lack of any property rights in the tissue taken lay at the heart of the California Supreme Court's final judgment in Moore's case. On the consent issue, it seemed abundantly clear that Moore had been lied to. He was told that extractions of further samples were essential to his treatment, which was not true: they were only crucial to development of the commercial cell line. His consent to the initial splenectomy was valid, but not the subsequent consents to the taking of further tissue.

But did Moore also have a right to be informed about possible commercial uses of the tissue? Is that a requirement for informed consent? Or must patients merely be told the medical facts rather than the commercial ones? That would mean they only had the right to know about risks, side effects and potential benefits of the procedure, rather than details of licensing agreements and clinicians' personal financial interests.

Because the standard for informed consent in the United States is—roughly speaking—what a reasonable patient would want to know, rather than the more paternalistic English guideline of what a reasonable doctor would reveal,[11] the California Supreme Court felt able to state that patients did need to know whether their doctors might be swayed by their own financial interests. 'The possibility that an interest extraneous to the patient's health has affected a physician's judgment is something that a reasonable patient would want to know in deciding whether to consent to a proposed course of treatment.'[12]

So in the final judgment, Moore was given the right to sue Golde for failure to obtain informed consent to the further procedures, and for breach of the doctor's fiduciary, good-faith duty to put his patient's interests above his own. That was as far as the Supreme Court was willing to go, even though the Court of Appeal had ruled in Moore's favour on the property issue. This fiduciary duty approach can only go so far: it can protect a patient from an individual physician who acts illicitly, but not from other researchers, employers or firms who don't owe the patient a fiduciary duty.[13] Moore was entitled to know about Golde's interests in the cell line when he was making his decision about whether to give or withhold consent to extraction of his tissues, but he wasn't entitled to any such interests himself.

However, the Supreme Court justices were radically divided in their opinions, and their disagreement goes to the core of the debate in this book. Do markets in human tissue—'body shopping'—debase our very humanity? Or is it only fair to allow those who donate tissue to profit from it as much as researchers and biotechnology companies do?

HOW MUCH WORK DOES IT TAKE TO MAKE A SPLEEN?

Although the dominant concern in the California Supreme Court's decision was preventing impediments to research from any further lawsuits like Moore's, at least one judge siding with the majority, Justice Arabian, did express his fears about 'the effect on human dignity of a marketplace in body parts'.[14] But in his dissent from the majority decision, Justice Broussard scoffed that body parts were *already* being valued in dollars and cents. Everyone stood to make a profit from Moore's tissue—except Moore.

> Far from elevating these biological materials above the market, the majority's decision simply bars *plaintiff* [Moore], the source of the cells, from obtaining the benefit of the cells' value, but permits *defendants* [Golde and the university hospital], who allegedly obtained the cells from plaintiff by improper means, to retain and exploit the full economic value of their ill-gotten gains free of their ordinary common law liability for conversion.[15]

In a view like Broussard's, 'human dignity' is just a smokescreen, unless it's backed up with concrete rights. Indeed, the majority in the Court of Appeal decision, which had found in Moore's favour, held that *not* to grant patients rights of control over their excised tissue would violate privacy and human dignity.[16] You could well argue that the real assault on human dignity comes from the exploitation of the vulnerable donor. Justice demands that both those who donate the raw tissue and those who use their skills to refine it should be rewarded proportionately, you might think, according to their labour. As Justice Mosk, another dissenting judge, wrote in his opinion:

> Recognizing a donor's property rights would prevent unjust enrichment by giving monetary rewards to the donor and researcher proportionate to the value of their respective contributions. Biotechnology depends on the contributions of both patients and researchers ... Failing to compensate the patient unjustly enriches the researcher because only the researcher's contribution is recognized.[17]

But how much of a contribution did Moore make to the original splenectomy, as opposed to the repeat visits to donate other tissue samples? Those weren't necessary for his treatment, but the first operation, to remove his spleen, was clinically essential. He would have had it anyway, as he acknowledged. So no additional work was entailed on his part. Neither can he be said to have put any work into making his own spleen.

Should Moore have been financially rewarded for undergoing an operation that benefited his health and entailed no actual labour? Was he wronged by not having been paid? The traditional common law position is 'no': Moore wasn't exploited, because he got what he wanted from the splenectomy, a possible cure for his leukaemia. That was where his true interests lay. Any value attached to the spleen after its extraction was purely fortuitous. It might even be unfair to reward Moore for having the good fortune of possessing particularly potent T-cells, by no dint of his own effort. (Of course we do allow people who chance to have particularly beautiful bodies and faces to reap rewards as models or actors, but they might be said to have put more effort into their appearance—dieting, undergoing plastic surgery, working out at the gym—than Moore, who had done nothing to produce his T-cells.)

The argument that we should be rewarded for our labour contribution, as made by Justice Mosk, may seem very much common sense, but like most common sense, it relies on the insights of long-dead philosophers. John Locke (1632–1704) is usually seen as the ultimate source of the belief that we own our bodies, and that therefore we have a rightful claim to own the results of 'mixing the labour' of our bodies with raw materials. If that is so, and if that labour is the primary source of an object's value, as Marx's conception of exploitation holds, then Justice Mosk is right.

But actually it's not so clear that we do own our bodies unreservedly. Self-ownership in the sense of ownership of the physical body isn't the crux of Locke's argument. We own our *actions*, as moral agents, but not necessarily our physical bodies. It's because I own my actions that I have a rightful claim to the resources or wealth produced by my actions, not because I own my body.

Although the conventional belief that we do own our bodies implicitly rests on Lockean foundations, in fact Locke never says that we have a property in our physical *bodies*, but rather that we have a property in our identities as *persons*. He is careful to distinguish between persons and bodies, and between the labour of our bodies and our bodies themselves, when he says: 'Every man has a property in his own person; this nobody has any right to but himself. The labour of his body and the work of his hands we may say are properly his.'[18]

What's mine, then, is the *labour* but not necessarily the *body*, the *work* but not the *hands*. We have a title to that with which we have 'mixed our labour' because our labour is the expression of our agency and status as persons, not just because the raw materials have touched our bodies. The connection isn't literally between our bodies and the hoe, flute or pen, but between our skills and the fruit, music or poem that flow from the labour in which we use those tools.

Further—and this is crucial for property in tissue, body parts or DNA—we don't have a property in that which we have *not* laboured to create. We don't own our bodies merely because 'we'—whoever that disembodied 'we' may be—inhabit them. In Locke's view, we don't own our bodies at all: God does, because he alone created them. Whether or not we accept Locke's belief in God, the point remains the same: if we only own that which we've worked to create, then we don't own our bodies. (One possible rejoinder—that we *do* work to maintain our bodies, by eating the right foods or going to the gym—quickly runs into problems. Does that mean that health fanatics own their bodies, but couch potatoes don't? Would the judges in a future *Moore* case have to work out a sliding scale to assess donors' claims to property in their tissue, according to the number of reps on the bicep press they do every week?)

Moore couldn't actually be assumed to own his spleen, then, because he hadn't worked to create it. In contrast, as the Supreme

Court majority was keen to point out, Golde and his fellow researchers *had* laboured to create the 'Mo' cell line, using their 'human ingenuity' and 'inventive effort'. In dissenting from the Court of Appeal decision that had found in Moore's favour on the conversion claim, Justice George had also likened Moore's spleen to mere 'raw material'. It had 'evolved into something of great value only through the unusual scientific expertise of the defendants, like unformed clay or stone transformed by the hands of a master sculptor into a valuable work of art'.[19]

So on the basis of how much work had gone into the valuable cell line, it's by no means clear that the decision in the *Moore* case was wrong, although Moore *had* put more effort into providing other forms of tissue after the splenectomy than the judges gave him credit for. But there could be many other reasons for calling the *Moore* decision ill-conceived. After all, the logic of the traditional common law position—that tissue taken from the body is 'no one's thing'—works against Golde as much as against Moore. If human tissue isn't the sort of thing that can be owned, patented or subjected to an action in conversion for recovery of property, then Golde, Quan, UCLA, Genetics Institute Inc. and Sandoz Pharmaceuticals had no right to the profits from Moore's cell line either. 'No one's thing' should mean precisely what it says.

That logic would have been devastating for Golde's side, of course. But as the dissenting judges at the Supreme Court level pointed out, the researchers and university would have been quick to press theft charges against someone who purloined the Moore cell line from their premises. So clearly they believed that human tissue in this form *was* something capable of being owned. It was hypocritical of them to deny that Moore could have at least some property rights in his tissue.

The California Supreme Court judges, however, decided to allow utilitarian arguments about overall good consequences for society to trump claims about an individual patient's rights. They clearly wanted to encourage biotechnology research, which they naïvely depicted as motivated by scientific progress alone. In that pretty picture, there is no mention of restrictive clauses imposed by drug companies on negative findings being published by researchers, or of patents taken out by biotechnology companies that hamper other researchers who can't pay the licence fee; yet these abuses exist, as later chapters of this book will demonstrate.

Instead, the majority opinion depicted an altruistic model of scientific progress, threatened by Moore's unseemly greed. 'At present,' the majority judges wrote, 'human cell lines are routinely copied and distributed to other researchers for experimental purposes, usually free of charge. This exchange of scientific materials will surely be compromised if each cell sample becomes the potential subject matter of a lawsuit.'

So why did the court think that Golde had taken out a patent? The whole point of patenting is to prevent free and unfettered use of a researcher's 'invention' without payment under licence. (Whether we can call a human cell line an 'invention' is an important question, which will also recur in a later chapter of this book.) Indeed, Lori Andrews, who was involved as an attorney in *Moore* and similar cases, has argued that the Supreme Court finding was a sort of self-denying prophecy. By providing researchers with every incentive to commodify tissues, immune from patients' lawsuits, it actually made them *less* likely to share samples and *slowed* biomedical research.[20]

Not only was the court's view politically blinkered, it's also logically contradictory. As the US academic lawyer James Boyle writes of the Moore decision in his influential book *Shamans, Software and Spleens*:

> On the one hand, property rights given to those whose bodies can be mined for valuable genetic information will hamstring research because property is inimical to the free exchange of information. On the other hand, property rights *must* be given to those who do the mining.[21]

The Court of Appeal had taken a more worldly-wise position when finding in Moore's favour: 'If this science has become science for profit, then we fail to see any justification for excluding the patient from participation in those profits.'[22] Misled by their jejune acceptance of an unduly altruistic picture of what goes on in scientific research, however, the Supreme Court judges in *Moore* gave almost all the spoils to one party: the researchers rather than the tissue donor.

After the California Supreme Court decision, Moore and the defendants settled, on the basis of breach of fiduciary duty, for an amount thought to range between $200,000 and $400,000—most of which vanished in legal fees.[23] Still trying to establish the point about ownership of his cell line, however, Moore appealed to the United States Supreme Court but failed to obtain a hearing. Both Golde and Moore are now dead.

DONORS OR DUPES?

Moore had succeeded only in establishing Golde's breach of the duties of a doctor, not his ownership rights in tissue taken from his body. He won on the informed consent and fiduciary duty grounds, but lost on the property issue. After Moore's death, a later case, *Greenberg*,[24] highlighted the inadequacy of relying on informed consent and fiduciary duty to protect tissue donors. By the time of the Greenberg case, twelve years after the final *Moore* decision, the court was much more alert to the question of unjust enrichment, but the donors still wound up feeling duped.

Daniel and Debbie Greenberg had lost two children to the rare genetic condition Canavan disease, a degenerative brain disorder primarily affecting families of Ashkenazi descent (Jews originally from Central and Eastern Europe). Approximately one in forty Ashkenazi Jews is a carrier of the recessive Canavan gene, named for the researcher Myrtelle Canavan, who first identified the condition in 1931. As with other recessive-gene-related disorders like cystic fibrosis, parents can carry the gene without knowing it, because they have no symptoms themselves. If two carriers have children, each child has a one in four chance of inheriting two recessive Canavan genes and displaying the condition. Children who only inherit the dominant non-Canavan gene from one parent and the Canavan gene from the other will be carriers themselves but won't be affected by the disorder.

Canavan disease is one of a group of inherited neurological disorders, including the better known Tay-Sachs disease, which impair growth of the myelin sheath, the 'white matter' of the brain that insulates the nerves. It's caused by an enzyme deficiency leading to a chemical imbalance that makes the 'white matter' spongy. Children with Canavan disease cannot crawl, walk, sit or talk; they may also develop seizures, paralysis and blindness. Most die before puberty; the condition is always fatal.[25]

After the deaths of their children, Jonathan and Amy, the Greenbergs contacted the genetic researcher Dr Reuben Matalon and convinced him that there was an urgent need to identify the genetic basis of Canavan disease. Had such a test been available at the time Jonathan was diagnosed with the fatal condition, the Greenbergs would not have had to endure the same trauma with Amy, conceived after Jonathan's diagnosis. The Greenbergs, determined to prevent other families from suffering as they had, founded a local branch of an organisation for parents of children with Tay-Sachs and related diseases, through which they met Dr Matalon.

Working collaboratively with Matalon over a period of thirteen years, the Greenbergs helped to build up a research bank containing tissue from other Canavan children, as well as their own son and daughter. Locating a total of over 160 other affected families, they arranged for samples of urine, blood, skin and even brain tissue to be donated to Matalon for his team's research into the genetic basis of Canavan disease. The Greenbergs established a database, the Canavan Registry, listing families with a history of the disease and specifying which tissues had been preserved at autopsy from those children. They also contributed substantially from their own purse and helped to raise funds for Canavan researchers employed by the Miami children's hospital, including Matalon himself. 'All the time we viewed it as a partnership,' David Greenberg said afterwards.[26]

They were mistaken. Without the Greenbergs' knowledge—just as in John Moore's case—the hospital at which Dr Matalon worked took out a comprehensive patent in 1997: US patent number 5,679,635, covering the gene coding for Canavan disease, diagnostic screening methods and kits for carrier and antenatal testing. Two years later, the hospital began to collect royalties from the patented genetic test, claiming that as a non-profit-making body—ironically enough—it needed to recoup its outlay on research. That was a piece of effrontery, since the initial seed money had actually been provided by parents like the Greenbergs and by the charitable Canavan Foundation, but standard pleading in medical research. It's a good reason to be sceptical of the claim that companies and researchers need to make profits in order to make good their losses, where start-up research funding is actually provided by the public purse.

By this time, the genetic test for Canavan disease was recommended by the American College of Obstetricians and Gynecologists for all Jewish women of Central and Eastern European descent—a sizeable market. Although the royalty was 'reasonably' priced at $12.50 per test, it was still too much for the Canavan Foundation, which had previously offered free testing. This, and other community-based screening programmes, had already helped to reduce the incidence of the disease by 90 per cent.

Now other parents would have to pay to find out whether their children might suffer from the condition, even though the Greenbergs and other Canavan parents had freely contributed the tissue, money and time that made the patent possible. Furthermore, the hospital attached a number of additional conditions, limiting the number of laboratories that could perform the tests and the number of tests to be performed annually. As a result, many laboratories stopped offering the diagnostic test altogether to parents trying to establish whether they were carriers.[27] Children who would suffer from Canavan disease were now being born unnecessarily, to a short life of suffering: precisely the outcome the Greenbergs had hoped to avoid for other families. In fact, it would have been better if they had never taken the initiative to get research started on the disease that killed their children. Now doctors were even barred, by the terms of the patent, from diagnosing Canavan disease through traditional methods *not* involving genetic testing.

In October 2000, the Canavan families and associated charities filed a lawsuit alleging breach of informed consent, breach of fiduciary duty, fraudulent concealment and unjust enrichment. Led by Daniel Greenberg and David Green, the plaintiffs (those bringing the action) also included the Canavan Foundation and the National Tay-Sachs and Allied Diseases Association. The hospital, together with Dr Matalon, were cited as respondents. The plaintiffs alleged that had they known that Matalon and the hospital intended to commercialise the discovery, they would either have withheld the samples and funding they provided, or found another researcher who had no such personal interests in mind. Such idealistic scientists do of course exist: for example, Francis Collins and his colleagues at the University of Michigan—the researchers who identified the gene coding for cystic fibrosis—decided to prohibit restrictive licensing agreements.

(Cystic fibrosis, another recessive genetic condition, is the most common inherited disorder affecting Caucasians.)

Three years later, in September 2003, the parties finally reached a confidential settlement, following the 2002 ruling of a court in Florida. Although that agreement was hardly an out-and-out victory for the Greenberg camp, which won only on the unjust enrichment claim, neither was it as clear-cut a victory for the forces of commercialisation of human tissue as *Moore.*

The other charges against Matalon and the hospital failed. Because not even the Greenbergs, still less the other families, had actually been Dr Matalon's *patients*, the court ruled that informed consent did not apply. This was not a one-to-one therapeutic relationship. Fiduciary duty, the grounds on which Moore had won, couldn't protect anyone but a patient; the Canavan parents weren't Matalon's patients. The individualistic slant of medical law, which does tend to focus narrowly on the doctor-patient dyad, is ill-equipped to deal with the wider sorts of research and corporate interests in cases like *Greenberg.*

But at least the starry-eyed view of altruistic science that shaped the *Moore* judgment was a little less prevalent by the time of *Greenberg.* In the latter case, it was quite clear that genetic patents could *impede* medical progress rather than facilitate it. If the Miami hospital insisted on a licence fee for other institutions wanting to perform diagnostic screening and clinical research, both Canavan families and scientific progress could and would suffer. The settlement exempted research doctors and scientists seeking a cure for Canavan disease from having to pay any royalty fee to the Miami hospital. A number of licensed laboratories were also granted an exemption from having to pay a fee for patients having the diagnostic test.

In return for these concessions, however, the plaintiffs had to agree not to challenge the hospital's ownership of the Canavan gene patent. The hospital was permitted to continue to license and collect royalty fees for clinical testing, except from exempted labs. It was little wonder that commentators generally reckoned that the Canavan decision still favoured researchers over tissue donors, just like the Moore case.[28] Although the hospital administrator who announced the settlement congratulated all parties on having united to fight the common enemy, Canavan disease, most of the spoils went to the hospital.[29]

So did most of the legal points in argument about whether there can be such a thing as property in the body. Again, the attempt to make a case for 'conversion' failed, because the *Greenberg* court still wouldn't accept that there was a property interest in body tissues or genetic material donated for research. If the plaintiffs didn't have a property in the donated tissues, then the respondents couldn't be guilty of interfering with that property. Following the majority judgment of the California Supreme Court, the judges in *Greenberg* resisted the notion that anyone other than the hospital or researchers could have a property interest in the Canavan patent and tissue bank.

Moore didn't labour to create his spleen, so on a Lockean basis there was no very good reason to say that he had a property in his tissue. But the Greenbergs and other Canavan families had put time, money and effort into creating the tissue bank on which Matalon's research crucially depended. Indeed, the whole thing was their idea: Matalon didn't possess any particular skills or interest in the genetic basis of Canavan disease, nor was he already working in the area. In fact, his group had already been refused funding from the National Institutes of Health because they had no track record in genetic research.

In 1999, the question about whether there can be property in the body had also been raised in the UK, when a sculptor, Kelly, made a brazen and rather cunning claim. Although he admitted to having taken preserved body parts from the Royal College of Surgeons without their consent, he said he had committed no theft of property—because there is no such thing in common law as property in the body. The College succeeded in recovering the body parts on the basis of a Lockean argument, that its members had invested their skills and labour in preserving them.[30] However, unlike members of the Royal College, Matalon possessed no particular expertise in the area where he was claiming property rights.

'So it turns out that the money and research skills were totally replaceable,' commented the US bioethicist Jon Merz. 'The only thing that was absolutely required in order to make the discovery was the participation of those families.'[31] In fact, the families' claim was so unusually strong, and their legal action so well organised, that it's hard to envision any future case resolving in favour of tissue donors and their families, if the Greenbergs and their allies lost.

The judge in *Greenberg* did recognise that the plaintiffs had contributed substantially enough to claim unjust enrichment; Matalon and the hospital didn't deserve *all* the spoils.

> The complaint has alleged more than just a donor-donee relationship for the purposes of an unjust enrichment claim. Rather, the facts paint a picture of a continuing research collaboration that involved Plaintiffs also investing time and significant resources in the race to isolate the Canavan gene. There, given the facts as alleged, the Court finds that Plaintiffs have sufficiently plied the requisite elements of an unjust enrichment claim.[32]

Unjust enrichment, in this sense, meant that one party had freely conferred a benefit on another, but that the recipient had gained financially from the benefit without paying for it. That, the court held, was inequitable—a form of exploitation, you might say. In the *Moore* case, Justice Mosk had likewise raised the issue of unjust enrichment in his dissent: 'Recognizing a donor's property rights would prevent unjust enrichment by giving monetary rewards to the donor and researcher proportionate to the value of their respective contributions.' But of course he was in a minority.

Here, in the *Greenberg* case, tissue donors might possibly see some basis for attempts to share in the proceeds, even without enjoying full-fledged property rights. In that sense, the decision in the Canavan case does represent an advance on the final opinion in *Moore*, but it still left the Greenbergs and other similarly situated patients and families without a firm right in the patent. That's very odd: if they were entitled to some of the spoils, it must have been on the same basis as Dr Matalon—of having put labour into the patent. If that effort was enough to ground a property interest for one side, the researchers, why not for the families?

Perhaps the issue just didn't arise in the same terms as in the earlier case. Moore was openly seeking a share of the profits from the 'Mo' cell line. The Canavan families and foundation primarily wanted control rather than money—the power to exempt diagnostic laboratories and researchers from having to pay a licence fee, to keep the database and the genetic sequence in the public domain. Although the plaintiffs did also file for damages of $75,000—fairly small beer by American standards—they were motivated benevolently, by the desire to assist the foundation. After all, that's why they had agreed to assist Dr Matalon in the first place, to help families at risk from Canavan disease. They assumed everyone else's motives would be as philanthropic as their own.[33]

No doubt that's why they felt sufficiently betrayed to start legal proceedings—just as Moore had, by Dr Golde's breach of fiduciary duty. In terms of the two possible senses of exploitation we explored in the previous chapter, they felt abused not by the financial imbalance in value between what they put in and what they got back, but by the way in which the dignity and value of their children's lives were ignored in favour of making money. Laurie Rosenow, an attorney on the plaintiffs' side, identified this motivation explicitly, stating that the Canavan families 'have charged their rights were violated because they were misused by researchers for financial gain'.[34] As David Greenberg described his feelings: 'What the hospital has done is a desecration of the good that has come from our children's short lives. I can't look at it any other way'.[35]

The bottom line is that the Canavan families gave freely of their time, money and beloved children's tissue—perhaps the ultimate gift. In a way, their loss was even greater than Moore's, and their trust in the altruistic purposes of the collaboration even more obviously abused. When a gift's purpose is misused this way, it negates the very value of giving. Yet throughout the biotechnology sector, it is increasingly assumed that donors of tissue or genetic data will give freely, without conditions or recompense. No such restrictions apply to the tissue banks or biotech firms, who can make full commercial use of the gift. As Robert Cook-Deegan of the Kennedy Institute of Ethics at Georgetown University has commented: 'We have a system where the research participants are treated as pure altruists but everyone else is treated as a pure capitalist.'[36]

Altruism is a good thing, of course, but one-way altruism smacks of exploitation and engenders mistrust, which actually undermines altruism. Presuming that patients will be donors, but making them feel like dupes, will kill the goose that lays the golden eggs. This psychological reality is what many utilitarian commentators miss when they argue that scientific progress must come first and that the consent of patients and families should simply be presumed.[37] That callous approach will inevitably backfire, as it did in the Greenberg case.

Unless a gift of tissue is shown proper respect—which doesn't just mean payment—appeals for tissue in the name of scientific progress will fall on cynical ears. 'You don't want research subjects to feel embittered and betrayed,' as one academic commentator remarks. 'In the long run, for research with human subjects to survive, those human subjects have to feel that they've been treated fairly.'[38] Surveys are already beginning to show that the public suspects that biotechnology is dominated by the profit motive, rather than concern for their welfare.[39]

Nearly forty years ago, the sociologist Richard Titmuss wrote an influential book, *The Gift Relationship*, which argued that altruistic systems of blood collection were both morally and practically superior to paid ones.[40] Contrasting what was then a largely paid system in the United States with an almost entirely voluntary one in the United Kingdom, Titmuss demonstrated that rates of blood-borne illness were lower in the UK (because underprivileged blood sellers in the US were more likely to have illnesses such as hepatitis) and that there was a national sense of pride in the UK blood collection system. That strong sense of identification with the National Health Service and the public blood donation system, he argued, both flowed from and helped to perpetuate 'the gift relationship'. Idealistic as that may sound, there was practical truth in it: my husband has so far donated his rare Group A rhesus negative blood forty-eight times, while both his mother and grandmother managed more than fifty donations each.

Titmuss is now thoroughly out of academic vogue, and the gift relationship is more honoured in the breach than the observance, although it is much trumpeted in official statements. Even on the political left, commentators argue that fractionation and commodification of blood products, even in the UK system, indicate that no purely altruistic system of tissue collection can long survive in late capitalism.[41] Further to the right, some writers openly and positively advocate the sale and purchase of stem cells and other tissues.[42]

Mixed private and public systems of organ and blood procurement now prevail in both the US and the UK. The squalid picture of 'ooze for booze' no longer applies in the US, in part because Titmuss's book did lead to reforms of the blood supply system. In the UK, however, NHS control over the blood supply system is mediated through private suppliers, and many recipients distrust any but their own blood after the scare over BSE ('mad cow' disease). The African-American writer Michelle Goodwin asserts that a similar mistrust prevails among black Americans, because of medical scandals such as the Tuskegee study, in which poor black men were allowed to die from untreated syphilis for fraudulent research purposes. She favours a modified market system in organs instead.[43]

Some institutional observers believe that whether tissue is sold for profit matters less than whether it is sold to government agencies or other rational bureaucracies, rather than to private buyers in a free-for-all.[44] Other more theoretical writers, including the enormously influential Jacques Derrida, scoff at the very notion of gift, 'deconstructing' it altogether.[45]

Gift requires counter-gift, as anthropologists widely recognise.[46] For the donor, giving is always about reciprocity, getting something back in return, even if the reciprocal satisfaction is a philanthropic sense of having contributed to the public good. Where a gift provokes further taking instead, and a sense of being duped—as in the Greenberg case—the whole structure of the gift relationship totters. One side cannot go on giving indefinitely while being taken baldly for granted. A genuine gift relationship also imposes responsibilities on the recipient.

If that notion were taken seriously, biotechnology firms, researchers and funders would urgently need to find new models of collaboration with tissue donors. In later chapters, we will look at some possible substitutes for an increasingly hollow and one-sided gift relationship. Because a few easily cultured cell lines like 'Mo' and 'HeLa' predominate in most laboratories, it would not actually be that difficult to give the originators of such cell lines, or their families, the control they seek.[47] And in most cases it is control rather than profit that they're after: Moore only sought a share in the profits because his attorneys could think of no other way to frame his ground-breaking case than as an action in conversion.

Before we cede ground to the demand that private organ sales should be allowed because donors are increasingly unwilling to give, we ought to try to give those who donate rather than sell their tissue the respect and control they deserve. 'Only connect,' as the novelist E.M. Forster wrote. Organ-donor families have organised reunions between recipients' and donors' kin, creating a modern form of the reciprocity that typifies gift.[48] Other forms of mutual respect and connection, this time between researchers and donors, might involve no more than periodically recontacting donors and allowing them to give or refuse consent to particular uses of their tissue, or to private versus public research collaborations. The 'conditional gift' approach is well worth trying and is already being tried in some quarters.[49] Why not? University endowments to provide a particular facility, or wills which impose conditions on beneficiaries, are well-established examples. Another possibility is joint patenting between patients and researchers, with profits flowing back into further research. The PXE International Foundation, set up by parents of children with the inherited disease *pseudoxanthoma elasticum*, provides an example of a possible middle way between pure altruism and pure capitalism.

So is the gift relationship still relevant? Actually, more so than ever, provided it's genuine. The altruistic psychology of gift still matters hugely to donors like the Greenbergs, and underpins the sense of injustice they felt when their gift was abused. The language of gift is now used to camouflage what is really exploitation, as when consent forms stipulate that tissue donors will have no further share in the profits from their tissue. Researchers, biotechnology companies and funding bodies certainly don't think the gift relationship is irrelevant; they do their very best to promote donors' belief in it, although it's a one-way gift relationship. What is really egg barter or sale, as we saw in Chapter One, is euphemistically called donation by the buyers. And in the next chapter, we will look at how commercial firms also employ the rhetoric of gift in a new form—the gift of umbilical cord blood.

3

'With love at Christmas—a set of stem cells'

> Christmas shopping for the unborn baby has never been easy. However, stem cell technology may have brought what is possibly this year's most original gift. For a mere £1,250, it is possible to harvest stem cells from the umbilical cord at birth and store them, frozen, for up to twenty-five years. 'Stem cells are not just for life—they're for Christmas', said Shamshad Ahmed, managing director of Smart Cells International, a company offering stem cell gift certificates as a new line this year. He has sold the idea to fifty customers so far—mainly grandparents who want their descendants to have access to stem cells' healing powers, in the event of illness or injury.[1]

A sum of £1,250 to ensure a child's future health: wouldn't every prospective parent who could possibly afford it want to store cord blood cells for the new baby? Stem cells extracted from cord blood can now be stored in private banks, in the hope that stem cell technologies will have advanced enough in the child's lifetime for them to come in handy. As an Australian father said of his decision to 'bank' with the firm Cryocite: 'I think it's quite clear that this technology is moving very quickly, and for not a huge amount of money, in fact quite a small amount of money, it's a good punt.'[2]

What is this technology, and what are its supposed benefits? It's been discovered that umbilical cord blood contains haematopoietic (blood-making) stem cells, similar to but fewer in number than those found in bone marrow, although even more flexible in their potential. These cells are what is called *pluri*potent, that is, capable of differentiating into a number of other tissues. Embryonic stem cells, which were first developed into separate cell lines by researchers at the University of Wisconsin in 1998, are the most flexible of all: genuinely *toti*potent, capable of making *any* other form of tissue. Umbilical cord blood comes a close second. Although the stem cells are present only in very low frequencies, it's thought they may have the potential to develop into many different types of tissue, including cartilage, fat, liver and heart cells as well as blood.[3] If the methods used in other stem technologies to encourage cell lines to grow are applied to cord blood, then possibly those cells could also be cultured into a wide range of tissues.

Bone marrow transplants are already used for certain blood diseases, such as leukaemia. Umbilical cord blood from other donors has also been used for similar purposes,[4] but more excitingly, it might additionally have the potential to be turned into a number of other forms of tissue matched to the donor herself. The potential advantage of stem cells taken from the cord blood at birth is that they would be tissue-matched to the baby, avoiding the problem of immune rejection that might arise from a transplant from another person. So, in theory—and it is still very much in theory—taking umbilical cord blood could provide the baby with a personal 'spare parts kit' for later life. And for people who object to embryonic stem cell research because they view it as killing embryos, umbilical cord blood appears to offer an ethically acceptable alternative.

In the United Kingdom, mothers can already bank cord blood in public repositories for transplantation, although most mothers may never have heard of that possibility, unless they happen to give birth in one of a few selected maternity units. The tissue types of both cord blood and bone marrow donors are available for matches anywhere in the world, including of course for the infant herself.

The public NHS Cord Blood Bank does also allow for 'directed donation' to at-risk families. Some doctors ask the public bank to store cord blood from infants born into families affected by a known genetic disease, where cord blood transplants can help. If these cells are compatible, they can be used in future for that child or a sibling. (The well-publicised phenomenon of the 'saviour sibling', deliberately conceived so that their cord blood can help an existing child, shows that compatible cord blood from siblings can be of definite benefit, for example in potentially fatal blood disorders.[5]) This altruistic cord blood banking service has been available since 1996, although it's currently restricted to four hospitals. What's new is the burgeoning of commercial cord blood banks.

When human tissue is an investment opportunity for Richard Branson, you know it's become just as much of an object of commerce as mobile phones, CDs or train tickets. Virgin's decision in February 2007 to set up a new business in umbilical cord blood banking is just another example of body shopping—although in this case, the tissue isn't sold to someone else, but back to the baby's parents. The Virgin Health Bank, a £10 million venture, is partly funded by the biotechnology venture capital fund Merlin Biosciences.

Branson's target is every couple expecting a baby—a huge market, at a cost of £1,500 for each baby. Four-fifths of the sample will be pooled for others' use and one-fifth retained for the baby—a minuscule amount, considering that the valuable cells only occur at very low frequencies. Capitalising, in every sense, on the fact that there are only four NHS cord blood banks and that their activities aren't well known, Branson plans to charge parents for the chance to be altruistic—which mothers who have access to an NHS cord blood bank can already do, without having to pay £1,500 for the privilege.

Branson is by no means the only corporate player, even if Virgin is the biggest one. Stem cells from umbilical cord blood aren't just for babies—they're for business. Commercial cord blood banks are not only on the rise in the United Kingdom, but also throughout the rest of Europe, as well as the United States. Often there is no bright line between their activities and those of their public counterparts. For example, the director of the non-profit Dusseldorf CB Bank is a scientific advisor and member of the board of directors of the profit-making firm Kourion Therapeutics AG, which estimates that the total cell therapy market in Germany will exceed $30 billion by the end of this decade. Kourion, in turn, was recently taken over by the US firm Viacell, parent company of the Viacord private cord blood bank.[6]

The potential market for umbilical cord blood banking is huge: every prospective mother, her partner and their own parents, in the classes and countries wealthy enough to afford it. Even developing countries like India are sprouting commercial banks for their wealthier citizens—at 59,500 rupees ($1,321), their very much wealthier citizens.[7]

Tinged with the glamour of stem cell technologies, cord blood banking apparently offers our children the proverbial elixir of youth. Supposedly, according to Catherine Waldby and Robert Mitchell, it:

> ... allows them to live in a double biological time. The body will age and change, lose its self-renewing power and succumb to illnesses of various kinds. The banked fragment, frozen and preserved from deterioration, can literally remake a crucial part of the account holder's body: the blood system.[8]

Here is an ancient incantation, even if the technology is modern. The myth of the infinitely regenerative body, the dream that every biological loss can be repaired—these are powerful hopes, especially when we hold them not on our own behalf, but for our children and children's children.

But if it's such a 'good punt', involving a comparatively small amount of money, why does the UK Royal College of Obstetricians and Gynaecologists generally advise against taking cord blood for commercial banking? The college was particularly worried about banking cord blood for premature babies, babies delivered by Caesarean section and twin births, but it was also troubled that in other cases, too, delivery room staff might be diverted from their main tasks at a critical moment. Overall, the College was very sceptical about whether the *speculative benefits* outweigh the *known risks.*

'TOTALLY SAFE AND HARMLESS'?

> The collection immediately after the birth of your baby is totally painless for mother and baby and does not present any risk. It is completely non invasive [sic]. The collection of the blood is only done AFTER the baby has been delivered. A small prick in the umbilical cord enables the blood to be collected ...[9]

> The collection of these precious stem cells is totally safe and harmless to both mother and newborn ...[10]

It's widely assumed that taking cord blood poses no threat to mother or baby, but that is in fact a false assumption, although the commercial cord blood banks do their very best to encourage it. Nor is the 'pay-off' from the 'punt' quite so attractive as the advocates of private banking claim. As the Royal College warns, 'despite the amount of interest in the field, the therapeutic role for such cells remains speculative'.[11] Let's begin with the risks to the mother, then look at risks to the baby, before evaluating the countervailing benefits.

Contrary to the cheerful impression given by commercial cord blood banks and echoed by a surprising number of otherwise well-informed ethical and legal scholars, the collection of umbilical cord blood normally takes place *during* childbirth, not afterwards. All but one of the private UK banks[12] prefer to extract the blood during the third stage of labour, between the delivery of the baby and that of the placenta, because more blood can be obtained that way.[12] As far as the mother is concerned, childbirth is not over after the baby has been delivered. In fact, the greatest risks to her lie in the third stage, since postpartum haemorrhage is the main cause of maternal death.

What exactly are these risks? The most obvious is that delivery room staff will be distracted from their main tasks. The first breath, foetal adaptation to the outside world and safe expulsion of the placenta are all complex processes. In the crucial and chancy third stage of labour, as thoughout childbirth, doctors' and midwives' primary duty of care is to the mother and baby, not to the priorities of a cord blood bank. They aren't employees of the blood bank, but professionals with a professional duty to perform. If parents or grandparents have signed a contract with a private blood bank, however, pressure may be brought on delivery room staff to fulfil it.

Because only a tiny fraction of the cells in cord blood are actually stem cells, capable of developing into many different forms of tissue, the private bank's interest lies in extracting as much cord blood as possible.[13] (In other words, they're not all 'magical', so to be sure you've got the 'charmed' ones, you need to take quite a lot.) Studies indicate that more blood, with higher stem cell counts, is obtained if the blood is collected while the placenta is still attached to the uterus, rather than after it's been expelled, and that there is also less possibility of infection.[14]

However, the mother needs a speedy and safe third stage of delivery, to minimise the risk of haemorrhage. There is some conflict between that requirement for the mother and the baby's need for the highest possible blood flow through the cord, although there the evidence is mixed. What seems quite clear is that the greatest conflict lies between the interests of either the mother *or* the baby, and that of the commercial cord blood bank. That conflict of interests exists even where public banking is concerned, but because there is less pressure to obtain as much blood as possible when the blood is going to be pooled, the conflict is less serious.

To see how greatly a birth involving extraction of cord blood differs from the 'usual' birth, let's sketch in the contours of a normal third stage of delivery. In an 'expectant management' third stage of labour, the baby remains attached to the umbilical cord, while blood continues to pulse for several minutes between the bodies of mother and child. The placenta would usually be delivered within thirty minutes to one hour and would then be separated from the cord. This process mimics that of other mammalian deliveries, where mother and baby lie still while waiting for the placenta to appear.

In 'active management' of the third stage, oxytoxic drugs are administered to hasten the separation of the placenta from the uterus, just as the baby's front shoulder appears. (Oxytocin in its natural form is a hormone secreted by the pituitary gland to initiate labour, stimulate contractions of the uterus and begin the process of lactation.) The baby takes a few breaths, the cord is clamped and cut soon thereafter, and the placenta is delivered by gentle pulling on the cord.

The greatest quantities of cord blood, however, are only obtained when the placenta is still in the uterus and the cord has been clamped immediately, even before the baby's first breath. This process would contravene current standards of good practice in the third stage under either expectant or active management. Clamping the cord before the baby's first breath obviously contradicts the more 'natural' form of delivery, expectant management. Active management includes clamping the cord, but *after* the baby has taken her first breaths.

Leaving the placenta attached to the wall of the uterus risks haemorrhage, the greatest cause of maternal death. That's why the balance of medical advice now favours routine administration of an oxytoxic drug, to stimulate contractions of the uterine muscles and ensure quick, clean delivery of the placenta.[15] Together with early clamping of the umbilical cord—while it's still pulsing—oxytoxic drug administration also reduces the length of the third stage of labour. No harm to the baby has been found to result.

The only harm from this procedure, in fact, is to the profits of the cord blood banks. To get the most blood possible, the umbilical cord has to be pulsing actively, meaning that the placenta still has to be attached to the womb. In a randomised clinical trial conducted by the private cord blood bank Eurocord, 100 collections made while the cord was still attached were compared with 100 taken after the placenta had separated. Significantly more blood was collected while the placenta was still attached to the uterine wall.[16]

In public banks, there is less pressure to maximise the donation, since cord blood is immunologically naïve: that is, lacking a strong response to tissue from another body, making pooled donations effective and allowing less perfect tissue matching for a transplant to succeed. A private bank, by contrast, will want to take as large a sample as possible, for 'security' and so that those benevolent grandparents feel that their Christmas present represents value for money.

Private blood banks have an obvious incentive to want more blood rather than less. Cryo-Care, for example, prides itself on obtaining not one but *two* lots of blood, 'for added security'. Whose security is served remains a moot point—certainly not the mother's. As a woman who has undergone childbirth twice herself, I wrote in an earlier article:

> The final stage of labour often sees the mother exhausted by pain and effort, only eager to conclude the business at hand by expelling the placenta, and to have her baby with her. She may well also have to undergo painful stitching of the perineum, if an episiotomy has been performed. How can it possibly be part of the doctor's duty of care to impose an additional burden on her by performing cord blood collection?[17]

From the viewpoint of risks to the mother, the evidence is limpidly clear: against taking cord blood. But you might argue that if a woman knowingly takes additional risks to benefit her child, that's wholly admirable. Unfortunately, it also looks as if the risks to the baby are greater than the benefits.

BENEFITS OR RISKS FOR THE BABY?

> Unimaginable possibilities … A once in a lifetime opportunity … Saving something that may conceivably save his or her life someday … Like freezing a spare immune system …[18]

> Cord blood stem cells are a miracle of nature that are only available once in a lifetime.[19]

> 'Our son Joseph is cured … The cord blood saved my son.'[20]

Commercial cord blood banks play up the speculative future benefits to the baby, but keep silent about the much better-established risks imposed by taking cord blood at the moment of birth. After all, presumably there is some physiological benefit in the blood still flowing from the placenta: it's 'nature's way'. Although the private banks' literature gives the impression that cord blood is merely a waste product, clamping at an early stage and preventing the full flow of blood may well deprive the newborn of something valuable. Delayed clamping can provide the infant with an additional 30 per cent blood volume and up to 60 per cent more red blood cells. That extra flow of cord blood results in additional iron stores, less chance of anaemia later in infancy, higher red blood cell flow to vital organs, better adaptation of the heart and lungs and raised likelihood of successful breastfeeding.[21]

The benefits are especially marked for premature babies. In a systematic review of seven studies, involving a total of 297 infants, delayed cord clamping for premature babies was found to improve their overall health, resulting in fewer transfusions for anaemia or low blood pressure.[22] One might assume that, for premature babies in particular, any blood removed is taken at a cost to the infant's health. Immediate clamping—which produces the most blood for banking—has also been reported to result in brain haemorrhage for premature infants, although it appears to be less harmful for full-term babies.[23] Some doctors doubt whether it is right to take any blood from the newborn because the long-term effects are still unknown; after all, cord blood transfusion is only a relatively recent phenomenon.[24]

Summing up, there are substantial risks to mother and infant from the strategy that produces the most blood for collection and storage.

Those to the mother are more obvious, but those to premature babies in particular are also well documented. Given that the first duty of a doctor is to 'do no harm', it's no wonder that the Royal College of Obstetricians and Gynaecologists advises its members to be cautious about cord blood collection.

But what about the possibility of therapeutic gains from cord blood to the baby or others? If the benefits are really as great as they're cracked up to be, they might outweigh the risks—and you might argue that the calculation of risks and benefits should be up to the mother. The problem is that mothers are being bombarded with highly misleading information.

A transfusion of the baby's own cord blood may not be the miracle cure touted in the banks' leaflets, so readily available in your local surgery or antenatal clinic. In fact, the consensus in the medical literature is that in the very unlikely event—about one in 20,000—of a cord blood transplant being advised, the patient would be better off using pooled blood from a public cord blood bank, rather than their own blood banked privately at birth. Those benevolent parents or grandparents may not be doing the new baby any service, and might also be putting both mother and child at additional risk during the delivery.

The private cord blood banks often list a very wide range of diseases and disorders supposedly treatable with cord blood stem cells. Before their usefulness in other non-blood-related conditions can be known, however, we need a much stronger evidence base and proper clinical trials. But that doesn't stop the commercial cord blood banks from making claims like: 'The current research indicates many areas where UCB [umbilical cord blood] stem cells may be used in everyday circumstances. Examples include cardiovascular disease, degenerative neurological conditions, tissue and organ engineering and diabetes.'[25] There's more than a whiff of wishful thinking here, and of preying on vulnerable people who suffer from these other conditions, or who know someone close to them who does.

It's true that there have recently been some encouraging but very preliminary results about the use of foetally derived stem cells in treating heart, spine and brain diseases.[26] Some of these studies were done in animals, so they're a long way from being proved or approved for humans. For many of the conditions listed by commercial banks, such as osteoporosis or immunodeficiencies, there is actually little or no evidence of any benefit.

Unless a family has a history of blood disorders, there is only a one in 20,000 chance that the infant will need her own blood during her first twenty years of life.[27] Blood disorders are the area in which cord blood transplants are best established, dating back to a case of Fanconi's anaemia in 1986.[28] (In this rare, genetically linked disease, bone marrow failure develops between the ages of five and ten, and the chances of survival are poor.) Cord blood continues to be most useful in cases like this, or in leukaemia and other haematological cancers, where it can lessen patients' dependence on bone marrow transplants.

But even in these cases, your own blood isn't necessarily the best. Contrary to intuition, the blood of others (an 'allogeneic' transplant from someone whose tissue matches that of the patient) may be clinically better for the patient than her own (an 'autologous' transplant).[29] Somehow, it seems that an allogeneic transplant increases the immune response and lessens the patient's chances of relapse after bone marrow transplantation.[30] There aren't any statistics yet for transplants involving cord blood rather than bone marrow, but it seems likely that the same logic would apply. If the source of the disorder is 'in the blood'—genetically based—the baby's own blood might even do more harm than good.[31]

That one child in 20,000 who will need a cord blood transplant is also unlikely to be able to rely just on the sample taken at birth. The 'once in a lifetime' opportunity might turn out to be a cruel illusion. A single sample is unlikely to be enough to treat any person who weighs over fifty kilogrammes, which would include most adults.[32] The median total stem cell yield of one cord blood unit is less than half the dose generally used in transplants. In the case of Branson's bank, which plans to store only one-fifth of the blood taken for the child's private use, the ratio is more like one-tenth (that is, one-fifth times one-half).

So parents who pay £1,500 to use the Branson bank will probably need to 'go public' in the end, supplementing the child's own blood with that of other tissue-matched donors, most likely from a public cord blood bank. Private health services are often accused of piggy-backing on public ones, using staff trained at public expense or facilities provided by the state. Private cord blood banking brings this point about free-riding home in a particularly acute clinical way. In addition to the need for doubling up with public blood, private banks such as Virgin want to use NHS staff to collect the blood—although the RCOG guidelines stipulate that collection must be done by a trained third party not involved in the delivery, so as not to divert the attention of doctors, midwives and nurses at a critical moment.

Let's recap: mothers in labour are being put at additional risk, for benefits to their babies which are largely speculative, in order to make profits for private cord blood banks. Privately banked blood is actually *less* clinically effective than pooled, publicly banked blood for the very few babies who will ever require a transplant. Getting your own blood isn't necessarily a good idea, particularly if the disorder being treated is 'in the blood' to begin with, as in a genetic condition, or in cases where pre-leukaemic mutations or leukaemia cells are already present in the cord blood of children who later develop that illness.[33] The medical case against private cord blood banking is largely damning. In fact, it's the twenty-first-century equivalent of Titmuss's argument about the 'gift relationship': for those who still want to give, altruistic donation is both clinically and ethically superior.

When Richard Branson announced the Virgin cord blood bank at a press conference in February 2007, he remarked: 'We are dealing with those [stem cells] taken from umbilical cord blood, which is normally discarded after a child is born. Using these cells as treatment presents no ethical issues.'[34] No ethical issues? Playing on expectant parents' sense of guilt if they haven't done everything possible for their baby could be seen as a form of emotional blackmail, when the evidence shows that cord blood extraction may harm both mother and baby and that the banked blood is unlikely to do any good. That, beyond doubt, is an ethical issue, particularly when we consider that only wealthier parents can afford to bank their infant's blood privately. That blood then becomes a sort of consumer good or venture capital, for the use of one person alone—unlike the public blood banks, open for all to use.

Public cord blood banking is still limited in scope, but it's growing. World-wide, by 2003, there were already over 70,000 units of placental blood stored in public banks, with an international search facility available to match blood samples with recipients. Even in the United

States, public banking has been established in twenty-two individual repositories such as the New York Blood Center and is to be extended into a more cohesive national system, with an appropriation of ten million dollars for a national system in the 2004 Federal budget. In France, public placental blood banks date back to the early 1990s, comprising traceable units which can be claimed back for a particular child's treatment. (The French national ethics committee disapproves of private banks, because they undermine social solidarity.)

Public banking has the additional advantage of being more representative of a good ethnic mix. Because tissue types vary between populations, and cord blood transfusions need to be tissue-matched to avoid rejection by the body's immune system, a public cord blood bank is more likely than a private one to serve minority ethnic groups well. Currently, about 40 per cent of donations to the UK National Cord Blood Bank (NCBB) come from minority ethnic groups, with the hospitals at which public donation is available having been chosen partly for their ethnic mix. That blood is available for tissue-matched recipients anywhere on the planet, of whatever national or ethnic origin. Private banks, of course, rely on ability to pay, and so discriminate in favour of wealthier majority ethnic populations and richer countries, even when, like the Virgin bank, they incorporate an element of sharing.

But if enough couples who can afford it decide to bank privately as a 'good punt', the stock of publicly donated blood will inevitably decline. Just as Titmuss argued, when the motivation of private gain infects a not-for-profit health system, altruism is undermined and public-spiritedness suffers. The irony is that it's not just the public health system that suffers in the cord blood case: so do those privileged mothers and babies who can afford to 'go private'.

WASTE NOT, WANT NOT

> ... [S]tem cells are available in large numbers from umbilical cord blood immediately after birth, something which in the past was simply discarded with the placenta.[35]

In the United States, there are twenty-two local public cord blood banks but no comparable national public bank to the UK model, although in the past few years Federal funding has been allocated to starting one, with the active support of American obstetricians.[36] As you might expect, US private banks strenuously reject regulation, appeal to their customers on the basis of 'biological insurance' and offer gimmicks such as a college savings plan as part of their package. The success of private cord blood banks in the United States has also been attributed to the frugal Puritan desire to avoid waste.[37] That argument—that cord blood is just junk—is frequently heard on the other side of the Atlantic. In announcing the Virgin cord blood bank, Richard Branson stressed that umbilical cord blood 'is normally discarded after a child is born'.

But is umbilical cord blood really just junk? We've already seen evidence that the baby needs it, especially the premature baby. Nor—contrary to claims like Branson's—is it simply extracted from material discarded after childbirth: it has to be taken during the third stage of labour, by an additional risky intervention. And of course there's a pivotal sense in which it isn't waste: it's valuable. Just as in the case of John Moore, it's hard to believe that anything so valuable could be 'junk', but it suits the interests of those who want to make profits from tissue to claim that it's worthless.

What the waste claim actually does is to mask the mother's rightful 'ownership' of the cord blood, making it appear to be something abandoned—open to the private cord blood bank to process and store, for a hefty fee. (That it is rightfully the mother's, rather than the baby's, will become clearer in the next section, but either way, it certainly isn't just abandoned, when the mother goes to the extra risk and trouble of having it extracted.) If cord blood is seen as abandoned tissue, the right to claim it seems to be open to the first comer. Waste tissue is common to all; the person who might once have had a particular claim in it has abandoned that right.

So if cord blood is presented as waste, it can all the more readily become the property of a private cord blood bank, by virtue of the 'effort' and 'skill' which the bank has put into storing it. Commercial cord blood banks in the United States often stipulate that if the annual storage fee isn't paid, the blood becomes the property of the bank. Private banks are already charging the mother for the privilege of giving her blood to the baby, at some risk to herself, as we have seen. It's adding insult to injury for them to deprive her of her own 'property' if she doesn't pay the storage fee.

In apparent—but false—contrast, a private UK bank, Cryo-Care, stresses that the stored blood remains the property of the parent. This is actually a meaningless reassurance. Whereas the US banks typically charge annual storage fees at a lower rate, Cryo-Care has already collected the full twenty-year storage fee in advance. Because the firm demands full payment up front, the question of what happens if payments lapse simply doesn't arise. True, Cryo-Care proudly advertises that it doesn't impose a further charge for retrieval of the blood, but this is hardly something to boast about. Private blood banks should not impose a charge for returning what was never theirs to begin with.

In both cases, private cord blood banks are actually assuming rights to which they're not entitled. If the test of a property right in something is putting labour and skill into it, doctors, nurses and midwives provide the skill and women in childbirth the labour. And in neither case should the cord blood be thought of as abandoned material, 'waste' to which the first-comer can lay a claim. Extracting cord blood requires a special separate procedure, not normally part of either 'expectant' or 'active' management of labour. Cord blood is *not* simply a normal waste product of labour that would otherwise be discarded.

We're back to the arguments in the *Moore* case, when those who stood to profit assiduously promoted the claim that the potentially valuable tissue was merely 'waste'. As we saw then, that strategy was more than a little hypocritical: if the tissue was only waste, why was there so much commercial interest in it? Just as Golde and his colleagues took out patents on Moore's supposedly valueless tissue, so firms are also starting to take out patents on cord blood—as well as reaping excellent rates of profit from the large 'customer base' they can expect.

As long ago as 1987, just after the first successful clinical use of cord blood to treat Fanconi's anaemia, and long before stem cell therapy was mooted, the Biocyte Corporation filed US patent number 5,004,681, for the cryo-preservation of neonatal and foetal blood and its therapeutic use involving haematopoietic cells.[38] Three subsequent patents were filed in 1988, 1990 and 1995, on the collection, processing and storage of cord blood, under the aegis of a new firm, PharmaStem Therapeutics, Inc., which also acquired the rights to the first patent and took out international patents in Europe and Japan. Following a challenge, the initial patents were overturned, based on the assertion that the company hadn't really made an original invention, but merely demonstrated that the cells could be frozen and thawed.

However, that didn't stop PharmaStem from suing five private cord blood banks that refused to pay royalties every time they collected a unit of cord blood. In 2003, the company succeeded in persuading a jury that the patents should be enforced against four out of five of the banks. The remaining bank had already settled with PharmaStem. These cases were still under appeal at the time of writing, along with a US Patent Office judgment revoking the patents on collection, processing and storage, but upholding the patent describing the therapeutic use of cord blood.

In June 2004, the firm sent out letters warning American obstetricians that they were infringing its patents if they co-operated with any of the four 'guilty' banks in extracting cord blood. The four banks must continue to pay royalties until the patent cases under appeal are finally settled. Public US banks would also be affected if PharmaStem succeeded in upholding its patent, but because they don't charge fees, they couldn't pass those costs on to the 'customer'. Although the PharmaStem patent was revoked for Europe by the European Patent Office in 2003, clinicians in Europe shouldn't be too confident until they see the outcome of the American cases.

WHOSE BLOOD IS IT ANYWAY?

> Researchers ... have cast doubt on such schemes but Mr Ahmed [of Smart Cells International] remains upbeat: 'Stem cells are a long-lasting insurance policy that has a once-only purchase date.'[39]

Insurance policy, discarded waste, mother's tissue, baby's private savings account: exactly what is the property status of umbilical cord blood? We've established that umbilical cord blood isn't just a discarded product which can be claimed by a private bank, but that does not settle the issue of who does have property rights in it—if anyone. Common law, as we've seen in the previous chapter, has always been very loath to allow people to have property rights in their bodies, however widespread the misconception that we do own our bodies.

There are good arguments in favour of recognising the labouring mother's entitlement to the cord blood. The Royal College of

Obstetricians and Gynaecologists statement on cord blood banking took the ground-breaking but authoritative position that cord blood is the mother's property, based on legal advice received by the College. If that view were more widely known, it would be harder for companies like PharmaStem to claim patents on cord blood.

> On one hand, it has been suggested that the cord blood sample is more likely to be the property of the child on the basis that it is developmentally, biologically and genetically part of the child. On the other, it might be proposed that it is more likely that the sample is the property of the mother ... Legal rights of property are not generally founded on genetic identity.[40]

The College found in favour of the second view, that the sample belongs to the mother. In terms of both law and physiology, this analysis has to be right. Our culture tends to privilege genetic identity—to believe that 'genes are us'—but that's irrelevant in this case.

The placenta is part of the mother's body throughout the third stage of labour, remaining attached to the wall of her uterus. When the cord blood produced by the placenta is extracted during that stage, then clearly that blood also comes from the mother and is hers. On the other hand, if the blood were taken after the placenta had been expelled from the mother's body, it might conceivably be seen as abandoned, unless she had expressed a desire to retain the afterbirth, as is done in some cultures. But that possibility is really irrelevant: most private cord blood banks wouldn't want the blood to be taken after the placenta has separated from the uterine wall, because less blood is produced that way.

Normally, the baby would receive from the mother all of the blood supplied through the conduit of the cord, until clamping occurs (in 'active' management of labour) or until the placenta is expelled naturally (in 'expectant' management). The mother is the donor of the blood in the usual situation, and the infant the recipient. That position holds even if the mother decides not to give all the cord blood to the baby at birth, but to store some of it, either privately or publicly. When cord blood is removed, it is taken from the maternal side of the clamp on the cord. It never enters the infant's body, or the portion of umbilical cord that remains on the infant's side. So there is no physiological reason to assume, as is often done, that cord blood 'belongs' to the infant. When cord blood is taken, a portion of that blood is donated by the mother to the cord blood bank, rather than to the infant. In private banking, it is donated for the baby's benefit, but the mother is always the donor: it is hers to give.

And she also puts effort into the process: as one childbirth manual says of labour, 'You've never worked so hard in your life.'[41] Although women's labour in childbirth is often seen as just a natural function, rather than conscious effort, there is little that is 'natural' about the extraction of cord blood. Women who decide to donate that blood on their baby's behalf are consciously and actively participating in a further intervention, which may prolong their labour and put them at some additional risk.

In our society we quite often reward risk-taking with ownership rights: that, after all, is the rationale for profits. It's supposedly because firms take risks, which may result in losses, that they are entitled to reap profits when they accrue. But in the case of private cord blood banks, the mother (and, to some extent, the baby) take risks, while the private firms make the profits and often claim the cord blood as their own property. If cord blood actually belongs to the mother, then private cord blood banks are charging her for the privilege of storing what is rightfully hers. The Virgin bank, which proposes to charge her for the privilege of donating it altruistically as well, takes that outrageous logic one step further.

Some commentators sceptical of private banks have dismissed them as akin to pawnbrokers, but at least a pawnbroker pays the client while the valuable object is kept in store. Here the client pays the pawnbroker. Perhaps a better analogy is a lock-up storage depot, although most people would blench at a contract stipulating that the depot could claim all their valuables if they missed a payment.

Arguably, those contracts are simply fraudulent. The bank has no prior rights over the cord blood, which is rightfully the mother's all along. She just doesn't realise it—because this is a new area of biotechnology and biolaw, because the private cord blood advertising leaflets are sometimes quite misleading and because even respected academics are confused about the issues. All these factors make cord blood collection prone to confusion of all sorts (not least on the yet-unanswered question of what happens if a private cord blood bank collapses).

Worse still, cord blood has been seized upon as a supposed cure-all in some cases where there is no evidence at all of any benefit.

CORD BLOOD, THE CURE-ALL?

James Logan, from Edinburgh, went blind overnight at the age of twenty-one, as a result of a rare and incurable genetic condition called Leber's optic atrophy. In February 2007, now aged forty-five, he underwent a 'revolutionary' therapy, based on umbilical cord blood stem cells of undisclosed origin. Dr Robert Troessel, who runs clinics in London, Antwerp and Rotterdam, claimed that the treatment would repair the damaged optic nerve and restore Logan's sight completely—for a mere £21,000.

Experts in Leber's atrophy have described Troessel's method as 'impossible biological gobbledygook'.[42] For the three people in every 100,000 who experience sight loss through Leber's atrophy, that loss is immediate, although the optic nerve continues to degenerate for another three months. During that 'active' phase, stem cell treatments might—might—conceivably have some effect. But in Logan's case, that phase was over twenty years ago.

Patrick Yu Wai Man, an expert on Leber's atrophy from the University of Newcastle , and his colleague, Philip Griffiths, a consultant ophthalmologist at the Royal Victoria Infirmary, have written:

> The bottom line is that there is currently no proven treatment for [Leber's atrophy]. We don't know the source of Dr Troessel's so-called 'stem cells', whether his preparation is safe for human injection or has received ethical approval from the relevant regulatory bodies. We are extremely concerned that James and possibly other patients might be subjecting themselves to unproven and possibly dangerous treatment.[43]

In other words, there's no medical reason to expect this cord blood 'cure' to work and plenty of reasons to think that vulnerable patients are being preyed upon.

One of those other patients was sixty-six-year-old Patricia Frost, whose progressive multiple sclerosis left her unable to feed, wash or dress herself. Patricia Frost claims that she felt instant improvements within an hour of her injections at the Preventative Medicine Clinic.

That supposed improvement ceased fairly quickly, 'and then when it stopped I was saying to my husband, "Oh, it's a con, it's a real con."'

Troessel is already under investigation by the Dutch health authorities and the General Medical Council in the United Kingdom. The Swiss firm with which he is associated, Advanced Cell Therapeutics (ACT), has already been ordered to cease its operations in Ireland by the Irish Medicines Board, for a similar cord blood-based treatment applied to neurological disorders. Denied a licence from that Board, ACT simply moved its operations offshore to international waters, where neither Irish nor European Community rules apply. Four hundred patients from Ireland and the UK underwent the procedures on the Swansea–Cork ferry in one month alone, May 2006. Even larger numbers of UK patients are thought to have undergone similar procedures at Troessel's Preventative Medicine Clinic in Rotterdam—the clinic Logan attended.

Executives of a related firm, Biomark International, which formerly provided the stem cells to ACT before it was closed in 2003 after an FDA investigation, have been indicted for fifty-one counts of fraud and other offences by a jury in Atlanta.[44] They stood accused of distributing stem cell treatment drugs which have no medical basis and of providing misleading information about the powers of stem cell treatments.

In May 2006, the BBC revealed that the stem cells supplied by ACT were not made for human clinical use and might contain animal material. Professor Neil Scolding, of Frenchay Hospital in Bristol, alleged that ACT refused to provide any scientific information on how the cell lines were prepared. The firm is accused of having bypassed research ethics committees in the Netherlands and of failing to carry out properly constituted clinical trials. In August 2006, a BBC Newsnight report claimed that ACT was selling research-grade stem cells for use in patients, although they had neither been screened for HIV nor used in anything but animal experiments.

The Preventative Medicine Clinic has since cancelled its ACT treatments, but at the time of the BBC report, it denied the allegations. The cord blood cells, it claimed, were:

> ... donated free by consenting parents in the First World ... They are certainly not animal cells, nor are they designated solely for animal

> studies. The cells supplied by ACT have certificates of analysis from accredited laboratories to prove their type, viability and purity... No recipient of ACT's therapies has ever reported adverse or negative side effects despite administering 736 treatments for over eighty conditions (primarily neurological) over a four-year period.[45]

But have they ever reported any benefits? Without properly conducted clinical trials, it's impossible to tell, but on the face of it, the method of administration—injecting cord blood stem cells through a subcutaneous incision—sounds thoroughly implausible. According to Colin McGuckin, professor of regenerative medicine at the University of Newcastle:

> If these cells are injected under the skin, they are extremely unlikely to have any effect as the immune system is designed specifically to reject foreign tissues. At most, I would expect them to cause nothing more than a rash.[46]

Just as diligent parents who want to do the best for their children are being targeted by the dubious tactics of some commercial cord blood banks, a similar sort of 'emotional blackmail' goes on at the other end of the umbilical cord stem cell supply chain. It's hard to avoid the conclusion that the vulnerability of patients like James Logan and Patricia Frost is being exploited by this type of 'stem cell tourism'—and that legitimate, productive stem cell research is also at risk of being besmirched. In the next chapter we'll further explore the mystique of stem cells, a powerful 'brand' in 'body shopping'.

4

Stem cells, Holy Grails and eggs on trees

A PIECE OF *SCIENCE* FICTION

In 2005, the Korean scientist Professor Hwang Woo-Suk published a paper in the highly respected journal *Science*, the tale of a successful quest to find the Holy Grail of stem cell research. Perhaps the proverbial man on the Clapham omnibus might not have recognised the Grail in the paper's title, 'Patient-specific embryonic stem cells derived from human SCNT blastocysts', but the biotechnology community certainly did.[1] What Hwang claimed to have created was a potential repair kit, tailored to any adult who could afford one: a personalised stem cell line capable of producing 'spare parts' which the body would recognise as its own tissue and which it could use to repair damaged organs or systems.

Here again, as in the umbilical cord blood story, we meet the old dreams of infinite regeneration, immortality and eternal youth—all in modern guise. Together with the consumer-friendly notion of a *personal* stem cell line—to match your personal MP3 player with your own personalised music tracks—those enticements create a powerful 'brand' indeed. Amid scenes of global media jubilation, Hwang promptly offered to franchise his team's expertise, creating a 'World Stem Cell Hub' with satellite laboratories in California and England. But events intervened, as they have a habit of doing.

Like umbilical cord blood, Hwang's technique would have avoided the problem of rejection, by which the body's immune system ferrets out and destroys invading foreign tissue. In theory, a single cell from my body could be turned into an immortal cell line matched to my own immune system—because the 'parent' cell was taken from my body in the first place. The genetic content of the fused cell would be identical to the genome of the person who gave the somatic (body) cell.

There, however, the similarity between cord blood stem cells and Hwang's technique ends. Cord blood stem cells are sufficiently flexible, or 'pluripotent', to produce a range of other tissues, if developed appropriately. Normal adult cells are already specialised, into skin, blood, bone or whatever; they have lost that flexibility. But they can, in theory, regain plasticity and develop into a variety of tissues by the somatic cell nuclear transfer procedure (SCNT).

In SCNT (sometimes known as 'therapeutic cloning') an *adult* cell is transferred into a human egg which has had its own nucleus removed. With a bit of encouragement, the remaining cytoplasm within the egg then 'reprogrammes' the transferred nucleus into a blastocyst (a very early stage embryo) and continues to divide like a naturally conceived embryo.[2] This was the procedure which Hwang and his colleagues claimed to have perfected, creating eleven new human stem cell lines from cloned human embryos.

A similar technique was used in 1997 to create Dolly the sheep, with the major difference that the reprogrammed sheep egg was allowed to complete its embryonic and foetal development and grow into a full-fledged lamb. In 'therapeutic' as opposed to reproductive cloning —quotation marks have been inserted because no therapies have actually resulted yet—the intention is not to create a cloned human being, but to produce stem cell lines for healing purposes. If there are ethical objections to therapeutic cloning, they have nothing to do with whether it's right or wrong to create a human clone. That is still a science-fiction scenario.

But actually Hwang's claim turned out to be a *Science*-fiction scenario. Although research performed in 2002 had shown that SCNT could partially restore immune function in immuno-deficient mice,[3] there were no success stories in humans before Hwang's 2005 paper (with the partial exception of a less complete paper he had published the year before). Yet the apparently rigorous refereeing process at *Science* couldn't be wrong—could it? Besides, Hwang's research protocol had been approved by not one but two research ethics committees.

Some bioethicists were sceptical of Hwang's claims from the start, because he claimed such improbable figures for the total numbers of eggs he and his colleagues had used. The media tended to ignore that question, concentrating instead on the promised cures, but feminist bioethicists in Korea and elsewhere were more alert to the issue of where the necessary eggs could have come from. With the rather similar 'Dolly' method, Professor Ian Wilmut and his team from the Roslin Institute near Edinburgh began with some 400 sheep ova (eggs), diminished to 267 after enucleation (removal of the nucleus). (Bearing in mind that ova are smaller than the head of a pin, taking out the nucleus is bound to be tricky.) Of those 267 eggs, into which a somatic cell from another sheep was introduced, only one developed into the 'success story' of Dolly. Yet Hwang claimed to have used fewer than two hundred eggs to produce not one but eleven stem cell lines—only eighteen eggs per line. Nor was he entirely consistent about who the donors of the eggs were.

In the excitement, however, most observers overlooked these qualms. Those who did express ethical doubts generally focused on whether it was permissible to 'kill' an early stage human embryo—albeit one created not through the normal fertilisation process, but rather by a form of cloning—and to create such an embryo purely with a view to its destruction, in order to create a stem cell line from it. Few observers (with the important exception of those alert to women's issues) seemed to want to ask where the eggs needed for somatic cell nuclear transfer were to come from, if this new technique were to become generally available. The newspapers, and even the scientific journals, were as silent on the issue as if human eggs grew on trees. Yet even extrapolating from Hwang's rates, rather than the twenty-times-greater wastage statistics in the Dolly technology, the activist Sarah Sexton calculated that developing a personalised stem cell repair kit for every diabetic in the United Kingdom would require between one-third and one-half of young British women to donate ova.[4] And that was just for one condition, diabetes, out of the many diseases for which somatic cell nuclear transfer was being touted as a hope of cure.

In the climate of adulation that followed Hwang's *Science* publications, tricky questions about the necessary supply of eggs were far from welcome. As Paik Young-Gyung of the activist organisation Korean WomenLink has put it: 'After Hwang's article was published in *Science*, he turned into a sacred figure in the South Korean public discourses.'[5] Paik notes that only 'pro-life' organisations questioned the research, focusing, as usual, on the status of the embryo, whereas WomenLink's concerns were the number of eggs that must have been needed, the validity of consent from the women involved and the conditions under which the eggs were obtained.

Together with ten other civic and feminist organisations, Korean WomenLink established a coalition called Solidarity for Biotechnology Watch (SBW) in July 2005, while Hwang was still firmly enthroned on his pedestal. This joint organisation kept the questions flowing about where Hwang had got his eggs from. 'It was the SBW which kept raising the ethical issues of Hwang Woo-Suk's research and questioning the validity of his articles at the same time. So if you argue that it was feminists who helped open up the scandal,' writes Paik, 'I think it is very much true to the case.'[6]

In particular, SBW brought to light troubling facts about the source of Hwang's eggs, which, it now transpired, had included junior researchers on his team. That was a serious breach of professional ethics and of the principle of voluntary consent. Voluntariness has been a key principle in research ethics since the Nuremberg Code, created after World War II to prevent any repetitions of Nazi experiments on concentration camp inmates. The subsequent Helsinki Declarations governing research ethics likewise stipulate that research subjects must never be forced or pressured to participate. If the Korean rumours were true, a grave affront to professional ethics would have occurred, a potential form of coercion. Junior researchers could well feel that their careers would be jeopardised if they refused to donate their eggs.

By November 2005, the results of SBW's investigations had apparently reached the ears of Hwang's research colleague Gerard Schatten, of the University of Pittsburgh. Schatten resigned from the team that month, giving as his reasons serious worries over whether the 'sourcing' of eggs had been conducted ethically. The University of Pittsburgh research committee had given permission for Schatten's collaboration with Hwang on the basis that no issues concerning human subjects were raised. Incredible as it may seem that the women 'donating' eggs were apparently not initially recognised as human subjects in the research, it does appear that Schatten became increasingly worried about their position.

Those concerns must have been grave, because Schatten also disclaimed any credit in the high-profile publications—worth their weight in research-funding-gold to any academic. Fool's gold, as it turned out—because with Schatten's resignation came the collapse of the entire Hwang enterprise, the revelation that the research claims were entirely fraudulent and Hwang's indictment on charges of fraud, embezzlement and violation of Korean laws against buying human eggs. Over half the women who 'gave' their eggs to Hwang had actually sold them to him.

Hwang hadn't managed to create a single successful stem cell line. Worse, he had wasted not something less than 200 eggs, but more like 2,200—over ten times that number—from 119 women. One woman had contributed forty-three eggs, implying sky-high dosages of ovarian stimulation—whose risks can be fatal. Fifteen to 20 per cent of Hwang's egg donors are now thought to have developed severe ovarian hyperstimulation syndrome,[7] as against a normal clinical rate of 0.5 to 5 per cent.

It would be too comfortable to dismiss Hwang's fraud as an unfortunate single occurrence that could never happen again, or at least, never happen in the West. One Korean analyst commented that Korean women are expected to sacrifice themselves for the common good,[8] reinforcing the old stereotype of subservient Oriental women. In fact, however, women in the West are also more likely to be organ donors than men. Living donors of a kidney were found to be disproportionately female in one US study—possibly because they were more likely than men to be asked, on the assumption that women should be altruistic.[9]

It won't do to assume that Eastern women are just more subservient, as a *Scientific American* commentator also implied in asserting that 'South Korea has a culture of egg donation for research'.[10] Assisted by Korean Womenlink and thirty-five other women's organisations, two women who gave eggs to Hwang have now started a lawsuit against the medical institutions and Korean government for failing to enforce the statutory protections they should have been afforded—which doesn't sound particularly subservient of them.

This lawsuit seeks compensation for failure to fully inform the women of possible side effects and for absence of informed consent.

Korea isn't the 'Wild West' of bioethics—not even the 'Wild East'. The Korean Law on Bioethics and Biosafety, which took effect on 1 January 2005, requires egg donation for both IVF and research purposes to be genuinely altruistic, with written informed consent from the donor after full explanation of the risks entailed. Trafficking of eggs through commercial agencies is prohibited, as is reproductive cloning, although 'therapeutic' cloning is allowed. These prohibitions are not so dissimilar to those that obtain in those European countries which permit SCNT stem cell research (currently the United Kingdom, Spain, Sweden and Belgium).

As the chair of a US Congressional subcommittee of investigation put it:

> Dr Hwang was not a rogue scientist operating on the fringes of his field with no oversight. He operated in an environment that proponents of cloning and embryo stem cell research would like to see adopted in the United States. Dr Hwang enjoyed the full support of his government, which vigorously promoted his research and funded it with tens of millions of dollars. Dr Hwang also enjoyed enormous popular support, and he had agreed to conduct his research under accepted ethical protocols. Dr Hwang's research was conducted with the approval of two separate Institutional Review Boards [local research ethics committees]. Nevertheless, Dr Hwang's actions represent the fulfilment of every warning dismissed by proponents of research cloning and embryonic stem cell research: thousands of eggs were obtained through payments and coercion; many women suffered terrible side-effects after they were not properly informed of the risks; not a single embryonic stem cell line was obtained for the tens of millions of dollars in government funds that were invested in the research; anxious patients were misled about the research potential.[11]

Although this is a comprehensive catalogue of 'the world's biggest scientific fraud of recent times',[12] it skates over one other development crucial to the theme of this book. Not only did Hwang improperly ask members of his research team to donate; he also paid other women an average of $1,400 for their eggs—body shopping. Those women constituted over half the total number of 'donors' (of course they were really sellers). And that trade was conducted on an international scale, through a for-profit firm, 'DNA-BANK', among other brokers.

DNA-BANK had been trafficking in eggs since 2001, amid plaudits for it in Korea as a model of innovative capitalism. The company began by recruiting egg donors for Japanese couples travelling to Korea for 'reproductive tourism' purposes—since surrogacy and egg donation are both illegal in Japan—but it soon 'diversified' into providing eggs, via the Internet, for research purposes. With links to hospitals and egg sellers in China and Malaysia and to a hospital directly represented in Hwang's research team, DNA-BANK had developed into a sizeable presence by the time its activities were unmasked at the time of Hwang's downfall.[13] (Some Korean newspapers were less incensed about the way in which the law had been flouted, or about the harm to the women from ovarian hyperstimulation for profit, than about the evils of selling Korean women's eggs to the old enemy, Japan.)

Like the commercial agencies supplying eggs for IVF, whose activities were explored in Chapter One, egg brokers who sell to researchers are springing up both in the Far East and in the United States. Since 2001, the Bedford Stem Cell Research Foundation near Boston has specialised in the provision of eggs for research rather than IVF. Although the number of its donors (twenty-three, as of July 2006, producing a total of 274 ova[14]) is still only minimal compared to the voracious demands of SCNT research, that position may well change. Enterprises like the Bedford Foundation are fewer in number so far than the profit-making IVF clinics which buy and sell women's eggs, but since the demand from the stem cell technologies is potentially even vaster, they could soon grow. Or else the institutions providing eggs for IVF might 'diversify' into sourcing them for research. That's exactly what happened with the Korean business DNA-BANK: it just converted its existing IVF-oriented networks in Japan, China and Malaysia into research egg procurement.

And since the genetic content of the enucleated egg is irrelevant in SCNT research, unlike IVF, why not go to the cheapest 'provider'? Poorer or non-white women's eggs are in little demand for IVF (at least in the US but not in Eastern Europe) because most couples who can afford fertility treatment are white and middle-class. Supposedly, disadvantaged women's eggs are thought undesirable because their 'uncertain genetic, behavioral and environmental health may also create a perceived "product liability" that would make their eggs unattractive to both IVF clinics and prospective recipients'.[15] But that rare instance in which poor women are *less* liable to exploitation doesn't apply to egg sourcing for research.

For the past six or seven years other researchers and I have been warning that we may eventually see a global trade in eggs for enucleation, with African or Indian women as the 'sources'.[16] It may take some time for that development to materialise, since most of the for-profit international egg brokers operate in the IVF field, where genetic content matters a very great deal—like the Cyprus and Ukraine clinics in Chapter One. But unless national and international governments wake up to the problem, it's no more unlikely than the development of specialist brokers who arrange clinical drug trials in cheaper countries, like many African nations. Those brokers have been operating for over ten years, serving the interests of the pharmaceutical firms.

Meanwhile, as the Korean commentator Paik Young-Gyung reminds us:

> Up to this point, we hear about the inappropriate acquisition or utilization of 'human ova'. Yet where are the 'people' who donated or sold ova and potentially experienced both physical side effects and a sense of betrayal after the scientific fraud was revealed? Why is nobody accountable for their suffering, legally, financially or even morally?[17]

STEM CELL RESEARCH: HYPE AND REALITY

The case of Hwang Woo-Suk may seem too easy a target. Wasn't he just a charlatan? Isn't it unfair to tar all stem cell research with the same brush? Of course, during the year or two when Hwang was the golden boy of stem cell research, many other researchers were quite content to bask in his reflected glory. Hwang did the 'therapeutic' cloning 'brand' a great deal of benefit then. It would be disingenuous of researchers in the same field to complain if the associations are unfavourable now.

Nevertheless, it's true that somatic cell nuclear transfer, or 'therapeutic' cloning, represents only a small fraction of current stem cell research initiatives. Perhaps because of the vast potential of a 'spare parts kit' for everyone who can afford one, however, it has attracted tremendous commercial and media interest, to the extent that many people wrongly believe that patient-matched cures for all sorts of diseases are just round the corner, if not already here.

It's been said that stem cell research encourages a view of the natural world as an artefact: 'to see the entire natural world, the human body along with it, as having the status only of material to be manipulated'.[18] By creating immortal stem cell lines touted as having the potential to reverse degradation and decay, we may even see ourselves as remoulding the biological universe. Government science policies have long tended to 'privilege the promissory', and stem cell research technology is the promissory technology *par excellence.*

The sociologist Melinda Cooper has suggested that Western free-market economies actively require this vision of an endlessly manipulable future.[19] She bases her claim on the work of Walter Benjamin, who wrote of capitalism: 'The gifts it dispenses emanate from a promissory future and forgo all anchorage in the past.'[20] Writing in the early twentieth century, Benjamin clearly had other 'gifts' in mind than those of modern-day biotechnology. But this idea of a 'promissory future' through regenerative medicine fits stem cell research remarkably well, just as it also does umbilical cord blood banking. What greater gift could be given than the life-sustaining alchemy of endlessly regenerative medicine? But are the promises of the promissory future merely hollow?

Doubts about the actual medical benefits of stem cell technologies aren't confined to non-scientists like Cooper, and they apply with particular force to 'therapeutic' cloning. Interviewed in *The Times*—not known as a hotbed of radicalism—a leading British stem cell researcher, Professor Austin Smith of the University of Cambridge, said that 'the promise of cloning for medical purposes has been oversold'. The emphasis given by the media to 'therapeutic' cloning (SCNT) has distorted the entire picture of stem cell research and given the public false hopes, in his view. 'There are real question marks about whether it has any utility at all,' Smith declared.[21]

In the United States, another factor in the 'hype' around therapeutic cloning has been the ban on Federal funding for new embryonic stem cell lines, enacted under George Bush in August 2001. Embryonic stem cell research, derived from surplus embryos originally created for IVF, involves a separate technique from therapeutic cloning, the method that Hwang claimed to have pioneered. Cell lines may be grown by isolating embryonic stem cells from the inner cell mass of a human blastocyst (a five-day-old embryo). These cells are cultured indefinitely with the help of fibroblast feeder layers. This is essentially the technique developed in 1998 by James Thomson and his colleagues at the University of Wisconsin, and subsequently pursued by a number of research teams.

Smith made it plain in his interview that he regarded embryonic stem cell (ES) research as far more promising than 'therapeutic' cloning (SCNT), as do many researchers. In the US context, however, progress in that form of research has been stymied by the religious right's abhorrence of 'murdering' embryos. (Since these embryos would never be implanted in any woman's womb, they wouldn't actually be killed by stem cell research: they would never have become living babies anyway. In our common law, personhood starts at birth, and since only persons can be murdered, the emotive language of 'murder' is just that: emotive.)

Of course we all want cures for diabetes, Parkinson's disease and the myriad other conditions for which future benefits have been claimed from stem cell research of either variety. But a note of caution is required—if not an entire symphony of notes of caution. This is not the first modern biotechnology whose speculative benefits have been portrayed as virtually boundless.

Ten years ago, gene therapy was the buzzword: the notion that individual disease-inducing genes in the body could be repaired by inserting normal DNA into cells to correct defects. But of almost seven hundred gene therapy trials approved in the United States, none has yet resulted in therapies approved for clinical use there or in Europe. Only a very few trials showed any efficacy at all, with safety issues remaining unresolved even in those. There have been two definite results, however—the deaths of eighteen-year-old Jesse Gelsinger in a 1999 gene therapy trial, and, eight years later, of thirty-six-year-old Jolee Mohr in another such trial. It later transpired that the principal investigator in Gelsinger's case had financial interests in the trial's outcome. Similarly, it has been alleged that the doctor who recruited patients into the trial in which Mohr died was paid by the firm financing the experiment, Targeted Genetics—as was the ethics board that was supposed to prevent such abuses. Connections of that sort are likely to become increasingly prevalent in stem cell research, too, because the US Federal ban is driving researchers into the arms of private funders.[22]

How similar is that dispiriting record to the actual state of stem cell research today? Admittedly, the Federal ban on funding might have slowed the pace of advance in the United States, but there has been copious private support, and sometimes substantial state funding: $3 billion in California alone. (However, the first of those Californian grants wasn't issued until March 2007, due to constitutional challenges from opponents of stem cell research.) Similarly in Europe, the European Commission funding framework programme VII doesn't allow money for research involving the destruction of embryos, but other sources have been found in those countries, such as the UK, whose laws do sanction stem cell research. Whether or not funding restrictions are to blame, stem cell researchers admit that they are 'not very close at all'[23] to clinical trials, much less to therapies ready for public use.

No one, at the time of writing, has yet managed to do what Hwang claimed to have done, although several research groups are trying to perfect the SCNT technique. There is also considerable scepticism about whether they will succeed, reflected in the reluctance of US venture capitalists to back this form of stem cell research. One of the only major 'players' in the US context is Advanced Cell Technology (ACT) of Massachusetts, which was reduced to a 'reverse buyout' in order to raise sufficient capital (a merger with a totally unrelated firm, a maker of Hopi Kachina dolls, which was already listed on the stock market and could thus offer ACT the listing it needed in order to raise capital).[24] Douglas Fambrough, a Boston venture capitalist, remarked: 'I'm not aware of any companies besides ACT doing it [somatic cell nuclear transfer].'[25]

'Ten years ahead there may be no need for ['therapeutic'] cloning, except in certain cases,' remarked Harry Moore of the Centre for Stem Cell Science at the University of Sheffield, UK.[26] It may well be more productive to tackle the problem of immune rejection directly, manipulating the immune system so that transplanted tissues and organs are less likely to be rejected. Moore thinks that there is much to be learned from pregnancy, the prime and primeval example of how the body can tolerate foreign tissue. If work on immune rejection makes progress, then the appeal of a personalised 'spare parts kit' would be much less. We saw in Chapter Three that the comparable promise of a personal repair kit through umbilical cord blood banking is also largely spurious, with better medical results from someone else's blood than the child's own.

Or it may be that other ways of deriving stem cells will eliminate the need for either embryos or eggs. Research published in *Nature* on 7 June 2007 by the Japanese scientist Shinya Yamanaka indicated that mouse skin cells could be reprogrammed into becoming pluripotent, by introducing into them four proteins which trigger the expression of other genes. If similar techniques could be developed for humans, and if ethical objections to tinkering with the genome could be overcome, proponents of this technology think that 'therapeutic cloning could be mothballed before it succeeds'.[27]

In the separate instance of embryonic stem cell (ES) research, things aren't that much further advanced: 'It looks like we're a ways off from being ready for prime time in man.'[28] Human trials of ES-derived cells will require rigorous safety precautions, partly because they will attract vivid media interest, but more profoundly because 'biological' therapies may possibly cause more unpredictable outcomes than standard 'chemical' drug trials.

In 2006, six young male volunteers at Northwick Park Hospital in London contracted sudden, massively life-threatening complications from a trial of a biological drug known as TGN1412. This experimental treatment was intended for multiple sclerosis, rheumatoid arthritis and other auto-immune diseases, in which the body's immune system reacts against itself. Although the drug had caused no adverse effects in monkeys, two of the men who took part in this 'Phase I' trial nearly died.

Standard research protocols set up what are called 'Phase I' trials to establish whether the drug being trialled may be toxic in humans. On the core principle in medical ethics of 'first do no harm', it is quite right that they should do so. But what sorts of toxicity might ES therapies produce, and how might that be minimised? These questions are rarely even considered in a media climate which promises untold magic from stem cell research.

Unlike the media, experts in the field often tend to be cautious about ES research: 'Long-term, I think there will be some therapeutic benefit, but I mean *really* long-term … I'm thinking ten years before we have an actual cure or benefit that's really tangible, and I'm being optimistic.'[29] Patient advocates, understandably, have welcomed plans by the US-based Geron Corporation to test neural derivatives of its embryonic stem cells in humans with spinal cord injuries. But some doctors, researchers and bioethicists have qualms about proceeding directly from favourable results in rodents (demonstrated in 2005 in a group of paralysed rats who were able partly to regain the ability to walk) to experiments on humans, without testing non-human primates first. Even the director of the University of California Institute for Stem Cell and Tissue Biology commented that without the precaution of primate trials: 'There is great potential for harm.'[30] And in the Northwick Park case, TGN1412 *had* been through primate trials.

Commenting on a proposal by StemCells, Inc. of California to begin a Phase I clinical trial on neural stem cell therapies in children with a fatal enzyme deficiency syndrome called Batten Disease, the US bioethicist David Magnus declared that there had to be a very strong prospect of therapeutic benefit if the research posed more than incremental risks to the patient. He added:

> My view is, for almost all these techniques, that they would not meet that standard for a prospect of benefit … When you have front-line, cutting-edge research, I'm very concerned that we are seeing a repeat of gene therapy—very thin, 'just-so' stories told about clinical benefit but with very little chance of things happening to benefit patients.[31]

Because body shopping is global, however, desperate patients are already travelling to countries with loose regulatory regimes to 'benefit' from untested stem cell therapies. As with the dubious clinics offering transfusions of umbilical cord blood which we met in Chapter Three, these agencies prey on the fears and hopes of patients who have heard tell of the powerful magic of stem cells. As described by Joshua Hare, a cardiologist at Johns Hopkins Medical School who is running a legitimate trial for adult stem cell treatment of heart attack victims: 'In Ecuador fetal stem cells, obtained in the Ukraine, are being used to treat patients from the US. There are cowboys who want to do this, and are going to do it.'[32]

So what can be done for such patients, if the outlook for stem cell research really is so unpromising? It's worth repeating that it's the techniques which have attracted the most hype—embryonic and somatic cell nuclear transfer stem cell research—that are actually furthest removed from clinical reality. *Adult* stem cells, although less totipotent or pluripotent than embryonic ones, have nevertheless yielded therapeutic benefits in some sixty diseases and conditions, including cardiac infarction (death of some of the heart tissue).[33] The cell with the greatest flexibility is the fertilised egg, which can produce any form of tissue. But stem cells found in specialised forms of tissue, such as bone marrow, retain a certain degree of this capability to produce other forms of cells.

Techniques involving adult stem cells should be distinguished from somatic cell nuclear transfer, in which an ordinary (non-stem) cell is taken from an adult and inserted into an enucleated egg. They constitute a third possible alternative to ES and SCNT research, without the ethical difficulties involving destruction of embryos or exploitation of women who provide eggs. Why are they less well known? Perhaps, speculatively, it's because embryonic stem cell research represents the Fountain of Youth to its supporters and the Slaughter of the Innocents to its detractors. Both inevitably attract media attention. 'Therapeutic' cloning, as we have seen, plays on the former myth, with the powerful additional resonance of 'personalised' medicine.

Whatever the mystical connotations of ES and SCNT research, the hard reality is that there is a no-ethical-holds-barred race among leading scientists and their nations to do what Hwang failed to do. The primacy of that motive was obvious in the decision in February 2007 by Britain's regulatory agency, the Human Fertilisation and Embryology Authority, to allow women to 'donate' eggs for stem cell research, which many felt was a foregone conclusion—whatever the risks. Given that the risks may be fatal, and that the benefits don't accrue to the women donating the eggs, opponents argued that the precautionary principle would dictate a moratorium or a ban on egg sourcing for research. Proponents of egg donation for somatic cell nuclear transfer stem cell research often argue that the risks are speculative, but as we've just seen, so are the benefits of that technique.

A RISKY ENDEAVOUR AND A *FAIT ACCOMPLI*

Of course human eggs don't just grow on trees, although you might be forgiven for thinking they do, from the usual media debates on stem cell technologies. Unlike sperm, they have to be extracted by surgical procedures. But that's only the third and last of three risk-laden stages in egg donation: shutting down the woman's ovaries, stimulating them to produce multiple follicles (rather than the single follicle usually produced in a cycle) and then—only then—extracting the resulting eggs. All this makes egg 'harvesting' an undeniably risky endeavour.

The usual drug used in the first process, shutting down the woman's own menstrual cycle, is leuprolide acetate (trade name Lupron) which has been reported as causing several side-effects, including arthralgia (severe non-inflammatory joint pain), dyspnoea (difficulty in breathing), chest pain, nausea, depression, dimness of vision, loss of pituitary function, hypertension, tachycardia (rapid beating of the heart), asthma, generalised oedema and abnormal liver function.[34] Irreversible losses of bone density, of up to 7.3 per cent of total bone, have also been reported.[35] The US Food and Drug Administration (FDA) has received many further reports of adverse effects that have not yet been investigated further. The FDA currently has on file over 6,000 complaints regarding Lupron, including twenty-five reported deaths.[36]

Next, the woman must undergo daily injections over a period of about ten days to stimulate her ovaries into producing extra eggs. This process may produce cysts, enlargement of the ovaries and fluid retention, with a potentially fatal outcome if ovarian hyperstimulation syndrome (OHSS) occurs. In that event, the ovaries swell and fluid builds up in the abdomen. Mild hyperstimulation will recede after the next menstrual period, but severe OHSS can cause blood clots, kidney failure, fluid in the lungs and shock.[37] A number of deaths from severe ovarian hyperstimulation syndrome have been documented, usually among women undergoing stimulation for IVF. Jacqueline Rushton, who died as a direct result of OHSS in Dublin in 2003, suffered a gradual deterioration of her organs, virtually all of which were slowly destroyed.[38] Temilola Akinbolagbe, a young woman who died in April 2005 in London, suffered a more sudden death from a massive heart attack linked directly to ovarian hyperstimulation syndrome.[39]

In March 2006, a study reported that ovarian hyperstimulation treatment in mice also results in abnormalities in their offspring. These effects include growth retardation, a delay in bone development and an eight-fold increase in a significant rib deformity associated with cancer. Animal studies can't always be generalised to humans (although it does often seem that there's a great rush to do just that when they're *favourable* to stem cell research, like the much-touted report, in June 2007, of embryonic stem cells produced from mouse skin). None the less, the authors concluded that their findings may have implications for the use of ovarian hyperstimulation treatments in women.[40]

Even in the absence of full-blown OHSS, ovarian stimulation in general has been linked in trials to pulmonary embolism, stroke, arterial occlusion and other life-threatening risks.[41] The incidence of the severe syndrome, which is known to have caused death, ranges between 0.5 and 5 per cent of cases.[42] However, it was estimated that a full 20 per cent of women 'donating' their eggs to Hwang suffered the full-blown syndrome. Where there is pressure to produce large numbers of eggs—as there was in Hwang's research—there may be a temptation to administer far higher dosages of stimulation than would be the norm in IVF.

These days, with the first 'test tube baby' nearing her thirtieth birthday, we tend to regard IVF as 'safe', but some commentators would like to think that egg donation for research is even safer. Both views are fallacious. Temilola Akinbolagbe and Jacqueline Rushton died of IVF-related ovarian hyperstimulation. And although doctors may have believed at one point that women donating eggs for research didn't suffer the complications to which patients undergoing IVF were exposed, a review of 1,000 cycles of egg donation found this was not true.[43]

In the third stage, the mature eggs are removed through a minor surgical procedure, laparoscopy or trans-vaginal ovarian aspiration. The donor is given painkillers or placed under intravenous sedation. Although conventional descriptions say the procedure should result in no more than 'mild pelvic discomfort', one egg donor said it felt 'like somebody punched you in the stomach'.[44]

Finally, we won't know the risks of premature menopause caused by the extraction of ova until the current generation of donors, now mostly in their twenties and thirties, reach the normal age of menopause. As it's generally accepted that a girl is born with a finite number of egg follicles, it stands to reason that taking ova out might produce early menopause, but it's just too soon to tell. (Research carried out in the 1950s by Solly Zuckerman showed that female animals, including humans, are effectively born with all their egg follicles. Work by the reproductive endocrinologist Jonathan Tilly in 2004 suggested that transplanted 'ovarian stem cells' in mouse bone marrow could migrate and repopulate sterile ovaries, but that research has been found invalid in subsequent studies, published in 2005 by researchers at Harvard and in 2007 by the scientists Lin Liu and David Keefe at the University of South Florida. No evidence of egg regeneration in women was found in comparable circumstances.)

You might think that the precautionary principle, 'too soon to tell' might translate into 'too soon to authorise egg donation'. But you would be wrong.

On 10 May 2006, Suzi Leather, then Chair of the UK Human Fertilisation and Embryology Authority, announced a public consultation on whether egg donation for research rather than IVF purposes should be approved by the HFEA. Her consultation statement appeared properly concerned about the balance of risks and benefits, but it also included a statement which gave grounds for concern: 'A paper was presented to the Authority today requesting a decision on the appropriateness of allowing altruistic egg donation—either from women who wish to donate purely for research or from women who are participating in an egg-sharing scheme and receiving discounted fertility treatment as a result.'[45]

The first set of concerns raised by this apparently mild statement of fact, nestled among the arguments for and against, involved procedural proprieties. If there was to be a fully open consultation, then how could a decision be handed down before the consultation finished? That would be a *fait accompli*. Why was Leather even mentioning that she had already received a request for a decision? She didn't need to receive any such request in order to announce a consultation on a topic of major importance. Was she perhaps thinking of allowing the authority to exempt the source of that request from the time frame of the consultation?

Secondly, why had Leather run together the two very separate sets of issues involved in purely altruistic donation and 'egg-sharing'? Although the HFEA had already allowed women undergoing IVF to donate a portion of their ova in exchange for reduced-cost treatment, those women were at least receiving some therapeutic benefit. In addition, women may be willing to 'share' when it gives another woman the chance of a baby, but there's no such direct tangible benefit to another person in the case of 'sharing' for stem cell research. As we've seen, any benefits from stem cell research are still a long way down the line.

Two months later, these qualms proved to be well grounded. On 28 July 2006, the HFEA gave permission to a team headed by Professor Alison Murdoch to pay women who agreed to 'share' their eggs a rebate of £1,200 towards the cost of their IVF treatment. Murdoch's group at the University of Newcastle (UK) had successfully created a SCNT blastocyst from excess IVF eggs, but not an entire cell line. They hoped to step into Hwang's shoes, it seemed, and the HFEA was prepared to make an exception in their favour—even before the consultation document had been sent out. That didn't occur until September 2006. It sounded very much as if the paper presented to the authority in May 2006 had been from the Murdoch team, and as if they weren't willing to wait.

Although it's illegal in the UK to pay for eggs used in either IVF or research, the £1,200 rebate was still a form of body shopping, making human eggs into a commodity. Under the euphemism of 'sharing', this transaction was plainly barter: offering a woman cheaper IVF in return for some of her eggs. Although Murdoch's team had already been granted permission in 2005 to ask IVF patients from whom more than twelve eggs had been collected to donate surplus eggs for research, they had collected only sixty-six ova over seven months. The gargantuan appetites of SCNT for human ova couldn't be sated so lightly. (Indeed, even while the consultation process was taking place, Murdoch's team and a team from another university also filed separate applications to be allowed to use enucleated animal eggs—so-called 'cybrids' or cytoplasmic hybrid embryos—as well as human ones.)

Not only does giving women a rebate on their IVF treatment constitute a form of body shopping; it also discriminates against less well-off women. Three-quarters of women in a similar situation in Belgium refused to share their eggs once their IVF treatment was fully funded by the public health service.[46] The financial incentive came first for the majority of Belgian women. There was no reason to think that the same wouldn't be true for British women, except that the UK's National Health Service was dragging its heels in implementing a national recommendation that women across the UK should receive three free IVF treatment cycles on their doctor's recommendation, eliminating patchy variation in public provision from one locality to another.

Egg 'sharing' also risks setting up a conflict of interest for clinicians, whose first duty must be to the patient. If doctors also feel they must obtain a set number of eggs for research—no matter how small (and small numbers are unlikely to satisfy the needs of stem cell researchers)—they're being pulled in two directions. They may possibly be tempted to administer excessive hormonal dosage regimes in order to collect enough eggs for research. Statistics from an HFEA inspection of the Newcastle Fertility Centre in 2004 already bear out this fear. A twenty-nine-year-old woman produced forty-four eggs after superovulatory hormonal treatment, while another young woman had twenty-nine eggs collected. It appears that twenty-three embryos were collected for research using the eggs obtained from these patients, while only eight were frozen for future use by the patients themselves.[47]

We saw in Chapter Three that the UK Royal College of Obstetricians and Gynaecologists was worried about a similar conflict of loyalties in the delivery room, in the case of umbilical cord blood. As I remarked in a press interview at the time the HFEA authorised egg 'sharing' at Newcastle: 'It's pulling the clinicians in two different ways. The primary objective has got to be to treat the individual patient, but if there is also a desire to harvest extra eggs, I think there could be two conflicting objectives.'[48]

The sense that the outcome was already a *fait accompli* was reinforced in September 2006 by the consultation document's title: *Donating eggs for research: safeguarding donors.* There was that magic appeal to altruism again, in the weasel word 'donation'

—and there, in the subtitle, under the equally charmed word 'safeguarding', lay a strong hint that there was no question about whether 'donation' would be allowed. The only issue was under what conditions.

That sense was reinforced by a further interim decision in January 2007 by the HFEA, again in favour of the Newcastle team, which was now allowed a 'temporary' licence to collect ova from 'non-patient' donors as well as egg sharers. It seemed that the Authority had been taken aback by the media storm over whether egg 'sharing' was a form of trading in tissue but thought the ethical issues stopped there. The consultation process had ended in December 2006, but it seemed the Newcastle team couldn't wait until the results were formally announced in February 2007.

So when the curtains were lifted on the outcome, the drama's *dénouement* had all the unpredictability of a badly scripted soap opera. Announcing the result, the HFEA's Chief Executive Angela McNab stated that:

> Today [20 February 2007] the Authority agreed to allow women to be able to donate their eggs to research projects, provided that there are strong safeguards in place to ensure that women are properly informed of the risks of the procedure and are properly protected from coercion ... Given that the medical risks for donating for research are no higher than for treatment, we have concluded that it is not for us to remove a woman's choice of how her donated eggs should be used.[49]

Let's pass over that specious concern for respecting 'a woman's choice', merely noting that what's driving the demand for eggs is not lengthy queues of women clamouring to donate, but high-stakes international competition among stem cell scientists. Let's focus instead on the argument that 'the medical risks for donating for research are no higher than for treatment'. If women willingly undergo the risks of ovarian hyperstimulation syndrome for one, why not for the other?

Within two weeks of the no-surprises-there announcement by the HFEA, the argument about comparable risks was pulverised by an authoritative study in the *Lancet*. The authors, a Dutch research team, offered convincing evidence that the best standard of treatment for IVF should *not* need to involve doses of hormones at the levels routinely used in gathering eggs for research. 'Mild' treatment strategies were proved in a clinical trial to produce as good a pregnancy rate as high-dosage therapies. Typically, one more cycle of treatment was required, but the 'take-home' baby rate was roughly the same, without the dangers.[50] The only threat lay to the commercial success of the IVF clinics, which are ranked according to success *per cycle* in the annual league tables produced by the HFEA.

If the 'mild' treatment strategy comes to be more widely adopted for IVF, then it becomes impossible to argue that women donating eggs for research happily submit themselves to exactly the same risks that they would if they needed IVF. Because the cloning technologies remain so bloatedly inefficient, they effectively require the multiple eggs that can only be produced by higher doses of hormones. In contrast, in the mild stimulation treatment regime for IVF, a woman ovulates more naturally, after a lower dose of artificial hormones.

Since high doses of hormones are *not* required for IVF, where the woman is at least deriving a therapeutic benefit, how can they possibly be justified when she is receiving no benefits at all? It is no part of the duties of a doctor to impose risks on members of the public who are not otherwise their patients, and who are receiving no such benefits. The only partial exception, as we've seen in discussing the Northwick Park case, is for participants in research trials. But are ova donors the equivalent of research participants?

In fact, women who donate ova aren't really subjects in a properly conducted research trial, but merely providers of raw materials for research.[51] No number of safeguards will change that position. At least the Northwick Park volunteers were monitored; Hwang's donors weren't. We still don't know the full extent of the damage done to them.

Nor are egg donors really equivalent to people who give solid organs, for example, a kidney. In that case, the recipient receives a direct benefit, the 'gift of life'—or at the very least, relief from the rigours and risks of dialysis. Given the speculative nature of somatic cell nuclear transfer stem cell research, it is by no means clear that any woman's eggs will produce any comparable benefit for a given individual, or indeed do any good at all.

TO PAY OR NOT TO PAY: IS THAT THE QUESTION?

Does it make any difference whether egg 'donors' are paid?—whether they are really egg sellers? The HFEA statement of February 2007 levelled considerable firepower at the notion that women would be paid £250 for their eggs, a story which had been leaked three days before in front-page headlines in the *Observer*. The authority punctiliously denied any such intention in its final statement, but we'll never know whether it hastily changed its position as a result of the *Observer*'s prediction that the decision would 'prompt a fierce backlash from leading figures in the medical world'.[52]

Admitting that women would be given £250 for 'expenses'—or, as some newspapers reported, £250 for forgone wages *and* an additional amount for expenses—the authority seemed unaware that women from Eastern Europe could easily be tempted by that sum. As we saw in the example of the Kiev clinic in Chapter One, the basic rate for egg sellers in some clinics is only three hundred dollars. If the 'expenses' were to be seen as wages, their full stinginess could best be understood in contrast with the number of hours women put into egg donation. The ethics committee of the American Society for Reproductive Medicine estimates that women spend a total of fifty-six hours per cycle of egg donation: the three-stage process involving injections to shut down the natural cycle, hormonal stimulation to produce extra eggs and extraction of the eggs under anaesthetic.[53] Even £250 in 'expenses' works out to less than the UK minimum wage of £5.35 an hour.

But is the real problem the low level of 'wages' for providing eggs? *Any* sum, according to some observers, is more than women deserve. 'Some will argue that an egg has no monetary value when it is just one of those ovulated each month by billions of women and that [sic] perishes unfertilized.'[54] No matter how vast the value placed on them by the hunger of stem cell research, in this view eggs are just a natural substance and women are merely fulfilling a natural function in ovulating. Of course that position ignores the labour women put into the patently *unnatural* process of egg extraction.

To a certain extent, the debate over whether women should be paid, and if so, how much, represents progress from the patronising and outdated view that egg donation is just a passive, natural process.

At least it recognises that women have a rightful property in their labour and its products. Isn't it fair to pay women who donate eggs for stem cell research a comparable amount to other research volunteers? The male Northwick Park participants received about £1,200 for their time and risk-taking. Only cultural prejudice, some might say, blinds us to the inequity of treating women differently, expecting them to sacrifice their time and tissue in order to benefit others. That, at least, is the argument put forward by the Korean-American ethicist Insoo Hyun, who chaired a task force exploring the issue for the International Society of Stem Cell Research. Hyun writes:

> The crux of the issue is this: if it is ethically and legally permissible for women to offer their oocytes for stem-cell research, and if it is acceptable to compensate healthy volunteers for their time, effort and inconvenience when undergoing comparable invasive procedures for research ... then there is a strong, presumptive reason to compensate healthy women who provide oocytes for basic research. Those who want to limit remuneration to women's direct expenses must rebut this presumption.[55]

One answer to Hyun's challenge has already been given: women giving their eggs don't enjoy the status of research participants, so parallels about paying research volunteers don't work. But there are deeper issues about whether payment prevents exploitation, or actually paves the way to it.

A commentator on Hyun's proposal countered, 'How can a fair level of compensation be set for risks that are virtually unknown?'[56] That's perfectly true; we don't have full information about some of the drugs involved in reprogramming the menstrual cycle, we don't have a very precise estimate of the incidence of OHSS, and we won't know the risks of premature menopause for another generation. Furthermore, we don't know enough about whether these drugs are carcinogenic. One large-scale study of 12,000 women who received ovarian stimulation drugs between 1965 and 1988 revealed no statistically significant increases in breast or ovarian cancer, but did find that women who had received them were 1.8 times more likely to develop uterine cancer.[57]

Should 'fair payment' include an element of compensation for a heightened risk of uterine cancer? Or is that invidious? Once a level of compensation is set, it will simply be assumed that if you've accepted it, you've accepted the risks as well. Money tends to close off the debate—but should it? One research study showed that higher payment levels don't function in research subjects' minds as a warning of higher risk.[58] If that is true, then we can't assume that subjects have accepted the higher risks just because they've accepted the money.

It would be worse than disingenuous for physicians who've extracted eggs to disclaim all responsibility for any adverse consequences, just because the women who provided those eggs had accepted payment. Eggs have to be extracted by medically trained personnel, and medically trained personnel owe their patients a duty of care. That's one reason why the US National Academies of Science and several US states (such as Massachusetts and California) have ruled out payment for egg donation in the research context. They're rolling back the commercialisation juggernaut, prohibiting payment for eggs given for research—even though a loophole in the US national organ procurement legislation has long permitted commercial trade in eggs for IVF.

On the other hand, payment for eggs to be used in research could be seen as more transparent than covert pressure of the sort that Hwang may have put on his junior researchers, or of the kind that families with over-optimistic expectations of stem cell research might exert on their female members to help 'save' a relative. It can be just as wrong to ask women to undergo these unknown risks for no payment as for money. In that sense, too, 'to pay or not to pay' isn't the question.

By asking for a share in the profits of the $3 billion cell line generated from his tissue, John Moore unfortunately and unknowingly set the tone for all the controversies about body shopping. Commentators on both the left and the right of the political gamut now typically cast the debate about human tissue in terms of a new version of Hamlet's question 'To be or not to be?'—'To pay or not to pay?' But Moore only framed his action in terms of conversion, recovery of property taken by another, because he was advised that the common law could find no other way to consider his claim. In the end, of course, he lost anyway, but the pattern was set. It's time to rethink this pecuniary approach, which has failed to protect those who most need protection. To pay or not to pay: that's *not* the question.

Too often, the idea of payment is being used to bypass crucial issues about exploitation of donors, on the grounds that if women have accepted payment for their eggs, they've also accepted the consequences. In this view, they have forfeited any right to further protection. But that argument just won't wash. In other areas, like health and safety at work, anyone but the most rabid free-marketeer would accept that the fact of being paid isn't the end of the debate. Employees accept their employers' wages, but they still retain rights to be protected from excessive risks, like being exposed to dangerous chemicals or being forced to work dangerously long hours. We don't think that it's paternalistic to enact those sorts of protections into law, even if workers have 'freely' chosen to take their employer's shilling. We just think that employers, like doctors, owe a certain duty of care, although we may disagree about the extent of that duty.

Well, then, how *can* women be protected from excessive risks in donating eggs for research? On the precautionary principle, and using the parallel of employment law, that should be the question, rather than whether and how much to pay them. The problem is that we just don't have enough evidence about those risks, in the long-term, to know how to protect women from them. To put the matter another way: of course women should be allowed to make an informed choice, but we don't have enough information to enable them to give a genuinely informed consent.

Unlike the UK, some countries, like Canada, have enacted a moratorium on donation of eggs for research, for precisely that reason. Not until we know more about the long-term effects of high-stimulation regimes or possible early menopause, and until we have much better evidence that SCNT research will be something other than a flash in the pan, can we create meaningful forms of legislative protection for egg donors. Concentrating on the payment issue shuts down those protection questions and lessens the pressure to get that evidence. That's one reason why 'to pay or not to pay' is not the question.

Another reason is that egg donors aren't really akin to organ donors, and so the arguments often advanced in favour of paid organ procurement don't apply to them. Advocates of paying for kidneys can set out firm figures about the shortfall between numbers of patients waiting for a kidney and numbers of kidneys donated altruistically. For example, Kieran Healy produces a dramatic table showing an implacably widening gap between the numbers of patients on the waiting list for organ donation in the United States and the numbers of cadaveric organs available. In 1988, the ratio of cadaver donors to patients on the waiting list was one to four—already a serious situation. By 2004, the ratio was more like one to eleven: eighty thousand people waiting for transplants, but only seven thousand cadaver donors.[59]

The argument is that allowing financial incentives for donation—either tax credits to the dead person's estate or direct payment to their beneficiaries—could directly lessen that acute shortage. One donor would more or less equate to one person off the waiting list, assuming tissue matching and successful immunosuppressive therapy. On a hard-headed basis, overlooking the 'yuck factor' of paying heirs for their loved one's organs, you could argue that we should be paying for organs because it will produce great social good.

Breaking those figures down by race adds a redistributive justice subtext to this argument. African-Americans have the highest death rate among all Americans on waiting lists for organs.[60] Registrants of colour represented nearly half of all those waiting for kidneys in 2003, with the majority of them African-Americans. The argument, then, is that the current system of altruistic donation fails to serve their interests, and thus indirectly perpetuates racial injustice. Whether or not you accept that argument, it's backed by definite figures about who needs a kidney and who could definitely benefit from paid donation, if payment increased supply.

But in the case of paid egg sale for somatic cell nuclear transfer research, the most that can be said is that certain groups of patients—say, diabetics—would benefit from SCNT-based treatments *if* they were successfully developed. That's quite a massive 'if'. One egg donor doesn't equate to one dialysis patient off the waiting list for a kidney; it may not even equate to any benefits at all.

If benefits do emerge from somatic cell nuclear transfer stem cell research, then it seems only right that those women who helped to make that progress possible should have some share in it. But again, 'to pay or not to pay' isn't the question. What Moore wanted, and what egg donors may well also demand, is recognition, gratitude and a degree of control—the status of collaborators in scientific progress, not just of raw material providers.

More and more attention is being paid to ways in which models like benefit-sharing or the drawing up of charitable trusts can give donors what they're really looking for. And the tide of expert bioethicists' opinion on further 'downstream' uses of tissue is swinging more in favour of donors, as against a blank cheque for researchers.[61] In the next chapter, on patents, we'll meet some examples of how such new models might work out in practice—along with some virulent examples of why those innovations are urgently required to prevent some of the grossest abuses in body shopping.

5

Genomes up for grabs: or, could Dr Frankenstein have patented his monster?

Before he was unmasked as a charlatan, Hwang Woo Suk had filed for a patent relating to the eleven tissue-matched stem cell lines he claimed to have created. Although the science turned out to be bogus, it seems possible that the patent applications might still be granted. But can a patent be valid even if the science behind it isn't? Although the common-sense reply would be 'no', in fact some commentators think the answer to whether Hwang's claim could succeed isn't at all obvious.[1] That shows how far patent law has travelled down the Yellow Brick Road towards an unreal and puffed-up Emerald City. In biotechnology patent law, we urgently need the equivalent of Toto pulling away the curtain, to reveal how flimsy the whole structure really is.

While the Great Oz was generally welcomed as a benevolent dictator[2]—even if he was really just a patent-elixir salesman from Kansas—the effects of monopoly patenting can be very pernicious indeed. By 2005, the number of patented human genes had increased to 4,270, representing 18 per cent of the entire human genome.[3] Of these four thousand patents, 63 per cent were held by private firms: one company alone, Sciona, holds patents on 2,500 genes.[4]

When these major biotechnology firms choose to do so, they can halt the onward march of science and impede clinical care—making nonsense of the usual accusation that those who campaign against body shopping are 'anti-progress'. Often, contrary to stereotype, it's actually the firms and researchers misusing patent law who stand most brazenly in the way of scientific progress, not the activists around the world who have opposed them. Why should we expect these corporations to be altruistic seekers after scientific truth? Their responsibility to their shareholders lies in making profits, not progress.

Even when they don't overtly impede other firms' researchers or clinicians' needs, corporate 'players' often design their genetic research strategy around which genes would be most profitable to patent, rather than which diseases most need cures. The two don't always mesh, not least because the 'worried western well' are more profitable.[5] For example, pharmaceutical firms have filed defensive patents on some genetic tests to undermine them *deliberately*, because tailored medications specifically for people with certain genomes would diminish the overall market for those drugs.[6] And yet the whole rationale of patents is that although they constitute temporary monopolies, they're crucial to the long-term flourishing of medicine and science.

One of the most scandalous cases of restrictive patenting has involved fees for diagnostic tests on two genes implicated in some breast cancers, BRCA1 and BRCA2, levied by the biotechnology firm holding patents on the genes, Myriad Genetics. Women with the 'wrong' version of these genes have a heightened risk of developing breast cancer (up to 85 per cent, as against the usual 12 per cent, although the genes account for only a minority of breast cancers). These women also run a greater risk of ovarian cancer.

In the United States, Myriad Genetics has a monopoly on all commercial testing for BRCA1 and BRCA2, with the full test costing up to $3,000.[7] In Canada, an alternative test has been available since 2003, despite threats from Myriad Genetics to launch a lawsuit against the Ontario Health Ministry, which developed its own tests for the same gene, using a different process. The Canadian test is both cheaper and faster, again undermining the argument that patents are good for patients.

A similar legal challenge from Myriad Genetics was quashed by the European Patent Court in 2005, revoking an earlier grant of patent in 2001. Myriad had argued for its rights to the 'isolated' version of the

BRCA1 gene, along with therapeutic applications and diagnostic tests involving the gene. This claim was eventually rejected, with the exception of a face-saving grant of patent on one single mutation of the gene, out of the thirty-four that Myriad Genetics had originally sought to control. The firm's claim regarding the BRCA2 gene was even more audacious. Although it had not discovered the gene's function, which was uncovered by Cancer Research UK in 1995, Myriad tried to muscle in when Cancer Research UK wanted to patent the gene. Cancer Research UK only aimed to protect its own rights to make the genetic test freely available for public health services.

In the United States, where its patents are still valid, Myriad has launched wide-ranging direct mailing shots to women, urging them to ask their doctors for a diagnostic test. This sort of scaremongering plays on patients' understandable confusion about the effect of the genes: although the vast majority of women with the BRCA1 gene will develop breast cancer, most breast cancers are not caused by the gene. So urging women to undergo an expensive genetic test for the sake of their peace of mind raises both false alarm and false hope: false alarm because the gene is comparatively rare, false hope because even if you test negative for the gene, you can still develop breast cancer.

But how can a firm like Myriad Genetics get a patent on a human gene in the first place? In an early landmark decision, *Diamond v. Chakrabarty*, the US Supreme Court held that patents could be issued on 'anything under the sun that is made by man'.[8] But how can it possibly be said that a gene is 'made by man'? And whatever happened to the idea that there is no such thing as property in the body? If one in five genes can be the subject of a patent, isn't the gene an obvious object of property?

CAN YOU TAKE OUT A PATENT ON LIFE?

Those who oppose genetic patenting often use the term 'patents on life' to describe what it is that they're resisting. It does seem extremely odd to think that genes—sometimes termed the building blocks of life—can be patented. Before looking at why the law allows genetic patents, however, we need to dispel one common myth: the idea that granting firms patents on the components of life is the same thing as giving them the power of life and death over our own individual lives.

That notion was tested in the 1994 Relaxin case, when opponents of a genetic patent argued that allowing the patent would be equivalent to allowing slavery. Relaxin is a natural hormone secreted in pregnant women by the corpus luteum (the progesterone-producing body which develops from the empty egg follicle); its role is to relax the uterus during childbirth. If drugs could be developed from Relaxin, they might be very useful in cases of difficult childbirth, including Caesarean deliveries.

The Howard Florey Institute in Australia, which had isolated and identified the DNA sequence generating the hormone, wanted to manufacture synthetic quantities of Relaxin for that purpose. To protect itself against commercial competitors with the same idea, it sought a patent. Having isolated the genetic sequence which codes for the hormone, the Florey Institute used recombinant DNA techniques to clone the gene, making it possible to produce synthetic Relaxin. What was to be patented wasn't the genetic sequence itself, as it occurred in any pregnant woman's body, but rather a copy of the gene produced in the laboratory.

The German Green Party mounted a challenge against the Florey Institute, arguing that the requested patent would contravene human dignity by using pregnancy to produce profits. The Greens also claimed that genetic patents constitute a modern form of slavery, involving the dismemberment of female tissue and its sale to profit-making companies. This, they said, contravened the general 'human right to self-determination'.[9]

Their argument failed. The European Patent Office rejected the objection that granting a patent would amount to a form of modern slavery over the pregnant women who had provided the genetic material to be patented. There was no risk, it said, of any one woman being forced to endure any form of bodily invasion by the patent-holder without her consent, still less of her becoming the patent-holder's slave. It wasn't literally the case that any one's body was being 'dismembered' in the course of producing a patentable genetic sequence, because that's not how the genetic sequence was produced.

It's important to avoid this confusion, since all too often the debate on the rights and wrongs of patenting the human genome slides into the unrelated non-question of whether it's right or wrong to own a human being.[10] The way in which Relaxin was patented is fairly typical of a genetic patent: it's issued not on the gene as it occurs in your body or my body, but on an identical version of that gene produced in the laboratory. Yet that's *exactly* what gives rise to the conceptual paradoxes in patenting.

On the one hand, no one actually owns the gene in your body or my body. It remains perfectly true that no one *can* own that; there is no such thing as property in the body in that sense. On the other hand, if I want to go for a diagnostic test to find out whether I have the abnormal version of the BRCA1 gene, I have to pay a fee to Myriad Genetics—at least if I want the test done in the United States. That seems massively inconsistent: if the patent isn't on the actual gene in my body, how can Myriad rightfully charge a fee to diagnose whether I have the faulty version of the gene?

There's a difficult tension here, which really pervades this whole book. When human biological materials are depersonalised through modern biotechnology—reduced into constituent parts or replicated in laboratories—they cease to be part of the person from whom they came. Human genes, in particular, begin to look more like abstract information rather than real tissue: a blueprint for how to build a human being, it's sometimes said.

As the Cambridge cultural geographer Bronwyn Parry writes, only recently have we learned to 'disaggregate' the human body in this fashion:

> New biotechnologies enable us, in other words, to extract genetic or biochemical material from living organisms, to process it in some way—by replicating, modifying or transforming it—and, in so doing, to produce from it further combinations of that information that might themselves prove to be commodifiable and marketable.[11]

These separated bits of biochemical material are much more like things than human beings, it can be argued. Slavery is not permitted because it's about owning human beings, who are not just things. If someone owns a patent on one of the genes occurring in my genome, however, that doesn't make me either a thing or a slave. (True, the US Patent Office has declared that a patent claim on the entire genome of any individual would violate the Thirteenth Amendment, which prohibits slavery.[12] But no such patent has yet been granted, although the artist Donna Rawlinson did file a patent application for an entity called 'Myself', consisting of her entire genome.[13])

Scientists and venture capitalists conventionally argue that, once replicated in the laboratory, disaggregated entities from the body exist 'out there'. In this argument, genes or genetic sequences can be seen as something apart from any one person's body, and as waiting to be 'discovered': a sort of virgin territory. But patent law rests on the argument that those same scientists or venture capitalists haven't just discovered a pre-existing law of nature; they've produced an invention, making something novel. Because it's their own invention, even if it's identical to the gene in its naturally occurring form, it's patentable.

No actual genes in any individual's body are owned when a patent is granted, and the kinds of rights granted are far from absolute. A patent only represents a time-limited monopoly over some aspects of management of the patented material, in exchange for free disclosure of information to the public at the outset of the patent term. For example, in the European Patent Office decision about the Harvard 'oncomouse', developed for cancer research, only the negative right to prevent others from using the 'invention' was awarded, not the positive right for Harvard researchers to use the mouse themselves.

But even that is just too much for those who oppose genetic patenting on a basis that we might call spiritual, for want of a better term.[14] If human genetic material partakes of the sacred—or is essential to human dignity, in a more secular formulation—then it makes no difference how small a segment is patented or how few powers the patent process actually conveys.

Yet why has the human genome taken on this iconic quality? Cord blood or ova might be expected to carry equal or greater emotional and symbolic freight: after all, they're crucially involved in the supposedly sacred process of human reproduction. And whereas ova can only be taken from their 'donor' through risky and painful processes, a DNA swab or a blood sample for genetic analysis can be separated from its 'source' without any physiological harm.

Even people who wouldn't actually term the genome 'holy' express concerns about genetic patenting. Somehow, the argument that no individual genome is being patented, so that no one is being enslaved in the process, fails to reassure many people that there's no problem.

Rather, the reverse is true: there's widespread dismay at the fact that the patent system dissociates the human source of the genetic material from the 'invention' itself. As the US academic lawyer Rebecca Eisenberg puts it, she's constantly asked the same question, usually posed with an air of righteous but genuine puzzlement: 'How can anyone possibly patent genes?'[15]

So what makes people so worried? Is it just that genetic patenting is 'unnatural'? That simply isn't a good enough explanation. All the new biotechnologies are 'unnatural'; for that matter, so is medicine itself. Couples who bank umbilical cord blood privately presumably realise that this technology is 'unnatural' but don't think 'unnatural' means 'bad'; they regard the possibility of cures being developed through stem cell technologies as very favourable indeed. All the processes by which patentable material is created are avowedly unnatural, and proud of it.

INVENTION OR DISCOVERY? THE CASE OF *DIAMOND V. CHAKRABARTY*

It's precisely *because* patentable material is 'unnatural' that it can be patented. The criteria for patenting include the crucial requirement of an 'inventive' or 'non-obvious' step. A related distinction is that the object of a patent shouldn't represent the *discovery* of something pre-existing, but rather an *invention.* European patent law explicitly excludes mere discoveries from patentability, while US law admits both discoveries and inventions but jibs at 'laws of nature and natural phenomena'.[16]

Yet how can a patent on a gene or genetic sequence possibly be said to represent an invention, rather than a discovery? As Eisenberg puts it, 'How can you patent genes?' Her answer is this:

> One cannot get a patent on a DNA sequence that would be infringed by someone who lives in a state of nature on Walden Pond, whose DNA continues to do the same thing it has done for generations in nature. But one can get a patent on DNA sequences in forms that only exist through the intervention of modern biotechnology.[17]

The argument widely accepted by national patent offices, and enshrined in the Trade-Related Intellectual Property Agreement

(TRIPS) governing international standards,[18] maintains that patents do *not* cover genes as discovered in their naturally occurring form. Instead a genetic patent demands the *inventive step* of creating genes artificially, by cloning and isolating them from the human body. While the original basis of the invention was a form of human tissue, that tissue has been reduced to the status of mere matter, no different from any other naturally occurring substance. The distinctively human element now lies not in the tissue itself, but rather in the inventive step by which technology transforms 'dumb' matter.

The further the object of a patent claim is from its natural state—the more 'man-made'—the more likely it is to fulfil the criterion of being an inventive step.[19] According to the influential holding in the US case of *Diamond v. Chakrabarty*, 'anything under the sun made by man' is patentable.[20] In that case, a form of a bacterium was patented—although you might well suppose that bacteria are natural rather than man-made.

Even in the United States, it was long assumed that scientific laws, or varieties of plants and animals, were indeed 'manifestations of nature [that must remain] free to all men [sic] and reserved exclusively to none'.[21] Patent protection couldn't be claimed for naturally occurring species or substances.

If you think back to the Lockean notion that property rights are grounded in labour, that makes sense. Patents are a form of property rights, and it's the labour of creating something new—like growing a crop where once there was just a fallow field—that gives rise to property rights.

However, you could also argue that discovering something new requires labour too. Many of the difficulties around patents—and body shopping more generally—concern the extent of labour that has to be put into something in order to ground a property right in it. As I asked in Chapter Two, concerning the case of John Moore, 'how much work does it take to make a spleen?'

From the point of view of the person from whom the original DNA is taken, does giving an innocuous cheek swab confer a right in the products developed from that cell line? That doesn't involve anything like as much time or risk as giving eggs for stem cell research. Somewhere in between comes John Moore, who was asked to return frequently after his original splenectomy to donate other tissue samples, none of which involved processes as risky as superovulation and egg extraction but which were laborious and time-consuming nonetheless.

From the viewpoint of the researcher, the same question arises: how much labour has to be put into a new development for a property right in it to be plausible? That was a major issue in the *Chakrabarty* case. In 1972, Ananda Chakrabarty, a researcher at the General Electric Corporation, filed a patent application for a genetically engineered bacterium that could be used to break down crude oil, as well as for the processes used to produce the bacterium. Initially, Chakrabarty's claim on the genetically engineered bacterium itself was rejected, on the grounds that bacteria are 'products of nature', not of human invention. As *Diamond v. Chakrabarty,* the case went all the way to the United States Supreme Court, where the claim was allowed. According to the justices, the bacterium was not a 'hitherto unknown natural phenomenon' but 'a product of human ingenuity'.[22]

Yet Chakrabarty himself admitted that 'I simply shuffled the genes—and changed bacteria that already existed.' Perhaps the Supreme Court thought Chakrabarty was just being modest. In the majority opinion, the researcher had applied his creative powers and scientific knowledge to produce a permanent physical transformation of a natural substance. On the Lockean argument, that might appear fair enough. Chakrabarty had put work and skill into transforming something natural into a product of his effort, you might think, and so he deserved a property right in that product.

However, the *Chakrabarty* decision contradicted an earlier case, *Funk Brothers Seed Co. v. Kalo Inoculant Co.*, in which genetic characteristics, as phenomena of nature, had been held to be inviolate.[23] That case involved naturally occurring bacteria in an artificial mixture: not far from *Chakrabarty*, which concerned natural genes in a man-made bacterium. 'At bottom, both inventions amounted to no more than the repackaging of naturally occurring characteristics.'[24] But in *Funk Brothers* the Court explicitly declared that an inventive step can't just involve repackaging an existing phenomenon of nature, which is really all that Chakrabarty did—by his own admission.

What made the difference? Possibly the state of the US economy. *Funk Brothers* dates back to 1947, a period of postwar industrial boom in the United States. In contrast, by the 1970s the US manufacturing sector had tumbled into decline, and great hopes were being pinned on the service and biotechnology sectors. Of course US Supreme Court judges are meant to be above political considerations, under a doctrine of 'strict constructionism'. But it's notable that in *Chakrabarty*, the majority of the Court refused to consider any arguments that didn't concern purely economic values—such as ethical or social arguments against patenting life. They implicitly accepted the common stereotype that patents are good for scientific research, and that scientific research is good for the economy. That may not be quite as crass as 'what's good for General Motors is good for the country', but it's not terribly sophisticated either.

The Court, in contradictory wise, held both that the social consequences of allowing life forms to be patented would be minimal, and that it wasn't competent to consider the social consequences—although by declaring them to be minimal, it had already considered them. And by ranking economic values above social ones, it had already made a value judgment, despite its contention that it wasn't the business of the Court to make value judgments.[25] The justices effectively elevated the man-made above the natural, and scientific skill above veneration for life. The legal scholar Richard Gold thinks that those implicit value judgments would have justified giving Dr Frankenstein patent rights:

> As in *Chakrabarty*, Frankenstein created his creature out of existing biological material. The monster itself, like the bacterium in *Chakrabarty*, did not exist in nature. While *Funk Brothers* would have held that life is too inviolate to be patented, under *Chakrabarty* the courts could very well hold that Frankenstein had exercised ingenuity in creating something that had never existed and might, therefore, have accorded Frankenstein a patent in his monster.[26]

Using its powers to interpret legislative intent, the US Supreme Court held that Congress had meant patent laws to be liberally interpreted, in the national economic interest.[27] Where natural phenomena can be transmuted into forms usable by industry, through an inventive step, they lose their inviolability. Naturally occurring substances remain in the public domain, as raw material for industry.

Although *Chakrabarty* didn't concern human tissue, that argument can be and has been extended to human tissue viewed as a raw material. We've already seen, in the Moore case, that human tissue can be used as just such a raw material, with patentable applications. The same would be true of patentable stem cell lines created by somatic cell nuclear transfer—if that technique is ever perfected—that rely on the raw material of women's eggs.

But even within the judiciary, there was substantial and well-reasoned dissent, just as in the Moore case. It's instructive, if depressing, to note that many of these influential decisions were originally very close things, but that because our common law relies on precedents in previous judgments, the majority opinions have controlled all subsequent developments in the United States, and, more indirectly, other common-law countries like the United Kingdom.

At the Court of Appeal level, one rung below the Supreme Court, four out of nine judges had found against Chakrabarty, arguing that US patent law was deliberately framed so as to forbid any one individual from 'securing a monopoly on living organisms, *no matter how produced or how used*'.[28] But a hybrid variety of seed can be patented; isn't a seed also a living organism? Perhaps it all depends on use, despite the warning 'no matter ... how used'. A wheat seed will germinate to produce a seedling, if given the right conditions, or it can be ground into bread, nourishing living organisms while not itself being allowed to live. Human or other animal genes, however, are another matter, so it's also unfortunate that the most influential decision in US patent law concerned a bacterium.

Equally ill-advised was the individualistic way in which the judiciary interpreted the notion of inventive contribution. Although research scientists generally work in teams and depend on the contributions of previous researchers, the patent system prefers to reward creative labour only when it's the product of a single 'inventor'. Chakrabarty himself pooh-poohed any notion that he was a brilliant lone genius. That stereotype is increasingly a fiction, particularly when so many patents are taken out on genes discovered through large-scale database scanning. The most blatant example was the way in which a private firm, Celera Genomics, applied for a daily round-up patent on whatever genetic sequences the publicly-funded Human Genome Project had uncovered that day.

Like it or not, scientific research these days is rarely conducted by lone Einsteins, Edisons or Faradays. Yet patents *are* readily granted on such communal 'inventions'; patent law is willing to overlook the original requirement of a single 'inventor'. By the same token, why can't collective rights be granted to those who supply the original human tissue?

It's sometimes said that it would be impractical and misguided to assign ongoing rights of any kind to groups of women who donate eggs for use in somatic cell nuclear transfer technologies, or to parents like the Greenberg group. But that argument just won't wash, if it's based on the contrast between rewarding single 'inventors' for their labour, but refusing to acknowledge similar work done by groups of tissue donors. A similar prejudice in favour of individual versus communal rights has often blocked Third World peoples from obtaining rights in plant varieties or herbal medicines developed through long communal traditions. Rarely can a single individual 'inventor' be named.[29]

Large-scale sequencing of entire genomes is less about identifying new chemical entities than about analysing patterns among genes. Most patents these days merely describe an association between a gene and a particular disease or condition, which looks much more like the discovery of a pre-existing correlation than a true invention.[30] Yet patent courts continue to regard DNA sequences primarily as chemical substances isolated and 'invented' by patent applicants.[31]

WHERE DO WE GO FROM HERE?

Since *Diamond v. Chakrabarty*, patent law has been interpreted—particularly in the United States—as allowing a biological organism to be patented as an invention, if it provides for a practical use of a discovery and not just the discovery 'as such'. Practical use may simply mean isolating the organism or genetic sequence from its natural surroundings, for example, in the human body. But that's usually *exactly* the case in which protection against 'body shopping' is most urgently required: when tissue is separated from the human body.[32]

As we've seen, the common law has traditionally viewed excised tissue as no longer belonging to the person from whom it was taken, and as open to all comers. What patent law does is to reinforce the rights of such 'prospectors' in the new Gold Rush over those of the patients from whom the tissue was taken. That affects *all* of us, as past, present or potential patients.

As Bronwyn Parry writes:

> As long as the material—such as a gene or biochemical substance—remains embedded within a whole organism, it is not considered to be patentable; however, once these elements are extracted from the body by any technical process, no matter how straightforward, they become eligible for patent protection. Advocates of patenting argued that to treat the isolation of a gene or compound as a mere discovery would fail to provide sufficient reward for the work entailed in achieving this result. Opponents argued that to treat isolated genes or compounds as patentable inventions when the sole contribution of the 'inventor' has been to reveal or discover an existing natural substance is to stretch the concept of 'invention' beyond reason.[33]

That's what's wrong with genetic patents: not that they reduce anyone to slavery, but that they do reinforce the powers of the already powerful. 'What makes you think you own your body?' becomes an even more unanswerable question, once the effect of patent law is taken into account. That law is being interpreted with increasing leniency: a survey of more than 1,100 US patent claims relating to human genes found that fully a third did not clearly meet the requirements of novelty, utility and inventiveness.[34] (Concern is now growing in the US Congress, where a draft bill prohibiting patents on human genetic material was tabled in February 2007 by Representatives Xavier Becerra [D-California] and Dave Weldon [R-Florida], but even if passed, the bill would not apply retroactively. Patents issued before it was enacted would continue to be valid.)

Particularly in the United States, patents are now routinely granted on living organisms of all sorts, as well as on components of human bodies. That may seem ironic, given the evangelical Right's defence of the 'right to life' at all costs. But such a right doesn't extend to protection against corporate patents, even on embryonic stem cell lines, which can be the rightful subject of a patent in US law.[35] A similarly liberal patent regime now obtains globally, with some 117 countries being automatically required, under TRIPS, to 'make patents available for all inventions, whether products or processes, in all fields of technology, provided that they are new, involve an inventive step and are capable of industrial application' (article 27.1).[36]

Some protection is afforded in European patent law by the doctrine of *ordre public*, under which an invention can't be patented if it's contrary to public morality. (The classic example would be a letter bomb.) Article 53 of the European Patent Convention also excludes as offensive to public morality the practices of human cloning, germline genetic modification, use of the human embryo for industrial or commercial purposes and processes for modifying animal genetic identity where harm outweighs benefit.[37] But the German Green Party failed to establish that *ordre public* applied in the Relaxin case, although there's greater hope in a recent European ruling against an attempt by the Wisconsin Alumni Research Foundation to take out a patent on human embryonic stem cells, which was refused on the grounds that granting a patent would indeed contravene public morality.[38]

Genetic patenting illustrates an important point about the new 'Gold Rush'. Human tissue *is* sometimes valuable in itself—so valuable that criminal rings will go to incredibly gruesome lengths to get hold of it, as in the case of Cooke's bones. The same is true of eggs for IVF from the 'right' kind of young women (the tall, blonde, intelligent, musical ones). But in many cases it's not the tissue itself that's the source of the value—rather, the patent line created from it. Whereas the theft of Cooke's bones incites us all to fear and fury, it's much harder to get exercised about a patent line. Yet that's where the legitimate big money lies, as opposed to the illicit activities of a few criminal rings, no matter how nefarious.

Like the wave and particle theory of light, genes, genetic sequences and other parts of genes seem to partake of two natures: informational and biological. Bronwyn Parry has suggested that biotechnology firms prefer to emphasise the informational, using metaphors such as the 'Book of Life' for the human genome, in order to position themselves as part of the vibrant new information economy. Perhaps they also do so because the courts interpreting patent law allow them to distance the genetic subject of the patent from the natural basis of it, to avoid the doctrine that phenomena of nature are sacred. The metaphor of the genome as information is two-edged, however: as we've seen, it actually implies that the 'book' is waiting to be read, rather than being written by the biotechnology firms. And that would make it a discovery, not an invention.

At the same time, patent law judgments contradict themselves by upholding restrictions on diagnostic testing for genes in individual human bodies, not in their isolated state produced by the inventive step. Genes predisposing to cystic fibrosis, breast cancer, Huntington's Disease and many other conditions have been successfully patented, drawing on the argument that they aren't present in the human body in their patented form.[39] But diagnostic tests assay the presence of those genes in actual human bodies; how can patent rights logically be upheld on diagnostic tests for those genes *in situ*? Yet so they are, with dire diagnostic and financial consequences for people who suspect they may be genetically susceptible to these conditions.

RESISTANCE IS *NOT* FUTILE: THE CASE OF TONGA

So must we just accept the great genome grab? The *Chakrabarty* decision, along with the logarithmic increase in genetic patents which it has sanctioned, might suggest so. But it's important not to assume that where the United States 'leads', the rest of the world must blindly follow. That would be both defeatist and condescending.

Elsewhere, there have been important challenges to the power of biotechnology corporations that seek to perform research leading to profitable patents, when that research conflicts with local beliefs and priorities. Two such instances of opposition to the patenting of life, the cases of Tonga and France, show that although genes and genomes may be up for grabs on a global level, the rise of commodification isn't irresistible. It has been, and continues to be, fought against.

In November 2000, the Australian firm Autogen announced to the Australian media an agreement with the Tongan Ministry of Health, to collect a bank of tissue samples for the purpose of genomic research into the causes of diabetes (well known for its high incidence, of about 14 per cent, in the Tongan population).[40] As the press announcement declared, the firm was attracted to the 'unique population resources of the Kingdom of Tonga'. Homogeneous indigenous populations possess an increasing appeal, not only in terms of research into the genetic basis of such conditions as diabetes, but also for pharmacogenomic and pharmacogenetic research (which involves learning how to tailor drug regimes on an individualised genetic basis).

If the tissue bank had resulted in important genetic discoveries, such as associations between particular gene sequences common to Tongans and the prevalence of diabetes, Autogen would almost certainly have sought to patent those sequences, any pharmaceuticals resulting from the work, and perhaps the processes by which those genes were isolated. So although the initial debate in Tonga revolved around the biobank, broader issues about genetic patenting and the meaning of the gene also came into play. That's where the greatest potential profits lay.

Although the Tongan public hadn't been informed of the initiative before the announcement in the Australian press, Autogen might have expected little resistance. It was offering several sorts of benefits: annual research funding for the Tongan Ministry of Health, royalties to the Tongan government from any commercially successful discoveries and the free provision of drugs arising from such discoveries to the people of Tonga. To be fair to Autogen, this package of shared benefits went beyond what many similar corporations would have offered.

But although the director of the Tonga Human Rights and Democracy Movement, Lopeti Senituli, had advocated similar benefits for indigenous peoples in a previous instance, when Smith Kline Beecham was pondering a bio-prospecting agreement for *plant* samples in Fiji, he was wholly opposed to the Tongan government's agreement with Autogen concerning *human* tissue—despite its apparently lucrative benefits. As Senituli put it:

> Existing intellectual property right laws favor those with the technology, the expertise and the capital. All we have is the raw material—our blood. We should not sell our children's blood so cheaply.[41]

It would be easy to dismiss this statement as a political war cry of dubious scientific accuracy. Of course the Tongans weren't literally being asked to sell their children's blood: the DNA samples to be taken were of renewable tissue in any case, and there was no theft of any individual's genome. But Senituli's position is mirrored in the views of many other peoples of the Global South, to whom benefit-sharing smacks of trinket exchange.[42]

Tongan society possesses sophisticated codes and moral systems, but they aren't rooted in private ownership. Even when sweetened by benefit-sharing, the Autogen proposals fell foul of those core Tongan values. The Tongan resistance movement raised the same issues about human dignity and the 'patenting of life' which occurred earlier in this chapter. Genes sum up who we are, what we have inherited from our ancestors and what we will pass on in turn to our descendants. (This notion of genes as held in a sort of venerable trust isn't restricted to indigenous peoples. Western writers have likewise asserted that genes and their DNA possess a 'sacred quality', which 'shares many characteristics with the immortal soul of Christianity'.[43]) As Senituli put it:

> The Tongan people in general still find it inconceivable that some person or Company or Government can own property rights over a human person's body or parts thereof. We speak of the human person as having 'ngeia', which means 'awe-inspiring, inspiring fear or wonder by its size or magnificence'. It also means 'dignity'. When we speak of 'ngeia o te tangata' we are referring to 'the dignity of the human person' derived from the Creator ... Therefore the human person should not be treated as a commodity, as something that can be exchanged for another, but always as a gift from the Creator.[44]

An additional value threatened by the Autogen proposal was *tapu o te tangata* (the sanctity of the person). In Polynesian belief systems, including the related culture of the Maori, the aim of a good life is to preserve and enhance *tapu*, keeping the self in a steady state of balance. As the eminent Maori cultural studies professor Hirini Moko Mead has written, Maori culture views one's personal *tapu* as the most important spiritual attribute of the individual: 'This attribute is inherited from the Maori parent and comes with the genes.'[45]

Actions by oneself or others that take away *tapu* are to be avoided. In the Polynesian context, it might well be thought that allowing others to take away one's genetic material is a violation of *tapu*, resulting in a diminution of the *tapu* available to one's descendants and affronting one's ancestors, who have striven to preserve their own *tapu* as a legacy.

Although learning for its own sake is highly esteemed in Polynesian cultures, research for principally financial gain does not necessarily share the same high value. On the other hand, if it could be known definitely that the proposed research might have lowered the high Tongan rate of diabetes or provided more effective therapies, the value of *tapu* might be displaced from its usual pre-eminent position. The countervailing value of *mauri* or life force could arguably be enhanced.

But there were additional hurdles. The Tongans particularly objected to Autogen's proposal that only individual informed consent was to be sought, in accordance with the dominant ethical model in Western genetic databanks. As Senituli demurred: 'The Tongan family, the bedrock of Tongan society, would have no say, even though the genetic material donated by individual members would reflect the family's genetic make-up.'[46]

Finally, in addition to their objections to violations of their central cultural beliefs, the Tongans also voiced pragmatic doubts. They cannily surmised that Autogen would reap rewards, such as higher share values and provision of venture capital from the pharmaceutical industry, as soon as the agreement with Tonga was announced—whether or not any therapies were eventually developed. By contrast, as Senituli said: 'the promised royalties from any therapeutics and the provision of those therapeutics free of charge to the Tongan people were, we felt, prefaced by a huge "IF"'.[47]

In the face of this opposition, in 2002 Autogen quietly dropped its proposed Tongan DNA biobank, announcing that it would conduct its research in Tasmania instead, but then disappearing from view altogether. So contrary to the usual cliché, 'resistance is futile', the Tongans' resistance had been extremely effective. That success undermines the argument, so prevalent these days, that 'We live in a capitalist society, so what do you expect?' Such cynicism is a self-fulfilling prophecy: if you don't believe resistance is possible, you won't resist. The Tongans believed the opposite.

Had Autogen acknowledged that harm had been done to Tongan values, regardless of the benefits offered, the resultant breakdown of negotiations might not have occurred. Possibly this seems an impossibly high price to exact of a Western company, and perhaps successful diabetes research might eventually have worked its way through to benefit the Tongan population. But then again it might not.

As we saw at the very beginning of this chapter, genetic patenting and data-mining can impede research as much as facilitate it. Where they do result in successful discoveries, those benefits often go to wealthy Westerners rather than the Third World subjects of the research. The enormous value that lies in rare genomes, and which genetic patenting seeks to tap, doesn't necessarily accrue to the 'owners' of those genomes. The Tongans were savvy enough to understand that fact, and to realise that even the proffered benefits didn't offset it.

Despite their communal emphasis, Maori and Polynesian values don't admit of the utilitarian calculus (according to which the greatest good of the greatest number, roughly speaking, outweighs all other considerations). Even if the benefit to be derived from the research were definite, there would still be qualms about sacrificing even a small part of some individuals' life force or *mauri* in order to benefit others.

By contrast, First World opponents of 'body shopping' inevitably have to fight the dominance of utilitarian thinking, both in academic bioethics[48] and in the popular press, particularly in the English-speaking countries. In Tonga much less credence was given to these utilitarian arguments, extolling, as if they were definite, the welfare and efficiency benefits of speculative biotechnologies, in which private market developers seek to extend property rights in tissue and genomes.

You might think that the communal values of the Tongans, and their emphasis on human dignity, couldn't prevail in a Western society. But there is also a major European country which has resisted body shopping in human tissue, including genetic patenting. That country is France, which sees itself as a bulwark against commercialisation and utilitarian Anglo-Saxon attitudes.

THE FRENCH DISCONNECTION

Attitudes and policy on 'body shopping' are consciously different in France: a phenomenon you might call 'The French Disconnection' (as opposed to the name of a certain film and a clothing chain). France has long refused to endorse the 1998 European biotechnology directive sanctioning most forms of patenting of the human genome, with an official governmental report stoutly maintaining that the directive would have to be renegotiated from square one before France would sign up.[49] For that stance, the country earned a formal censure from the European Commission in July 2004. But France continues to refuse to sign the directive, on the grounds that it makes life into a commodity—which is actually very similar to the Tongans' objections in the Autogen case.

In 2000, the French Minister of Justice, Elisabeth Guigou, flatly declared that human genetic patents violate French ethical principles.[50] French national documents and commissions frequently, if somewhat sanctimoniously, present their views as firmly principled and altruistic, as against those of the laxly 'pragmatic' or 'utilitarian' Anglo-Saxon countries. The particular principle nearest to the French *cœur* is the insistence that the body is the person, and is therefore inviolable. Branded a 'taboo' by the French political scientist Dominique Memmi,[51] this tenet isn't just abstract; it can actually influence practical policy.

In 1994, the French government blocked a research collaboration between the American biotechnology firm Millennium Pharmaceuticals and a leading genomics laboratory, *Centre d'Étude du Polymorphisme Humain* (CEPH), on the grounds that 'French DNA' should not be given away. That sounds remarkably like Senituli's insistence that 'We should not sell our children's blood so cheaply.' As in Tonga, the external biotechnology firm had its eye on a unique genetic resource: in this case, an extensive pool of genetic data collected by CEPH from a large number of French families. And again, coincidentally, their interest lay in families with a high rate of diabetes. CEPH also had a prior collaboration with a non-profit organisation for families with muscular dystrophy, *Association Française contre les Myopathies* (AFM), resulting in a large and valuable pool of genetic data on another genetically linked condition.

While the French government had initially approved of proposals to use these resources for the collaboration between CEPH and Millennium Pharmaceuticals, in 1994 it changed its mind. Unlike in Tonga, where the data-mining arrangement was concluded between the government and Autogen, in the French instance it was the government, rather than activists, which stepped in to scupper the deal. Genetic data from French families belonged to the nation as a whole, the government claimed, and couldn't be appropriated by a foreign firm.[52]

Materials freely given as public gifts by patients and their relations should not simply be transformed into private capital resources, the government insisted. (In addition, there were also allegations of insider dealing and conflicts of interest: the director of CEPH, Daniel Cohen, was also a founder of Millennium Pharmaceuticals.) So whereas in the United States during this period the biotechnology landscape was being massively reshaped by large amounts of venture capital investment in research-industry collaborations, aided by a change in the climate of legislation, France consciously resisted that model at the highest levels of government.

Like the Tongan case, the example of 'French DNA' could just seem negative: rejection of the supposed benefits a deal would have offered, rather than successful negotiation of a deal that would have genuinely done both parties good. As the American author of *French DNA*, Paul Rabinow, puts it: 'the invocation of "genetic patrimony" fits snugly with the main symbols of French bioethics: menace, integrity, identity'.[53] But on the other hand, is that necessarily a bad thing? Preserving national integrity and identity against what may well be a genuine menace from globalised 'body shopping' isn't just some strange obsession; it's a practical necessity.

The notion that the body is the person, and therefore sacrosanct, has also been valiantly defended for well over twenty years by the French national ethics committee, *Comité Consultatif National d'Éthique* (CCNE). CCNE has consistently taken the strongest possible stand against the commodification of human tissue, which it calls 'an intolerable disrespect for the person, a radical violation of our law, a threat of decay to our entire civilisation'.[54] As its first chairman, Professor Jean Bernard, insisted: 'Ethics has no worse enemy than money'.[55] Lucien Sève, a long-standing member, described his colleagues' stance as radically hostile to the body-shopping mentality:

> If there is one characteristic that genuinely typifies what I call 'bioethics *à la française*', as developed and promulgated by the CCNE, it is this: intransigence towards all spirit of lucre in research, all corruption in biomedicine, all commodification of the human body.[56]

In 1984, the year of its establishment, the CCNE's very first opinion presciently denounced the commercial use of foetal tissue, at a time when awareness of that possibility in other countries was minimal.[57]

Later, in its 1991 opinion 27, 'That the human genome should not be used for commercial purposes', the CCNE set out two principles 'to which the Committee attaches the most fundamental importance'. One of these is 'the inviolable principle that the human body cannot be put to commercial use'.[58] The other is the argument that the human genome is the common property of humanity as a whole, translated in French as *patrimoine de l'humanité*. In the Anglophone world, this argument is more often heard from activists who oppose genetic patenting, whereas in France it bears the stamp and sanction of an august committee of the great and good.

Likewise, CCNE opinion number 21, 'That the human body should not be used for commercial purposes' (1990), proudly declared that:

> The view of French law on this problem is clear. It does not accept that the human body should be used for commercial purposes. The body is not an object and cannot be used as such; for instance, blood and organs are not for sale, a position which is rarely encountered elsewhere.

(The UK's blood donation system, likewise based on free donation, doesn't get a mention, but then this statement manifests a fair bit of Gallic exceptionalism.) In the French civil code,[59] as restated by this CCNE opinion, 'the human body or one of its components cannot be the object of a contract'. So women can't sell their eggs, for example. The reasoning behind this prohibition invokes the principles of human dignity and non-exploitation:

> For instance, an organ, such as the kidney, cannot be sold by the person to whom it belongs and, even if it is donated free of charge, cannot be sold by a third party, however much the would-be recipient or his entourage insists on it. Such insistence may be tantamount to blackmailing dependent individuals, for example prison inmates or misused minorities. Human dignity is at stake if financial gain becomes the result of physical weakness, however temporary.[60]

Conceiving of organ sale as an issue of *social* justice and power relations typifies the French style. *Individual* consent from the kidney seller doesn't suffice to outweigh questions about protecting the vulnerable.[61] And the vulnerable person is seen not as the recipient—as in Fabre's analysis, although Fabre is herself French—but as the seller.

Another core principle is mutuality. A very telling example of the difference between Anglo-Saxon and French perspectives in this regard can be found in the CCNE's opinion number 74 (2002), 'Umbilical cord blood banks for autologous use or for research'. Rather than posing the question in terms of benefits to individual babies or the choice of individual couples, the CCNE opinion condemns the private banking of cord blood for autologous use as a breach of social solidarity:

> Preserving placental blood for the child itself strikes a solitary and restrictive note in contrast with the implicit solidarity of donation. It amounts to putting away in a bank as a precaution, as a biological preventive investment, as biological insurance ... There is a major divergence between the concept of preservation for the child decided by parents and that of solidarity with the rest of society.[62]

Solidarity is linked to gift, whose centrality in French bioethics dates back to the two world wars. Before World War I, blood was paid for, but now the system is based on altruism and gift—although some commentators fear that the effect of European Community membership will be to reinstate a market system.[63] During the 1980s, an intense national debate over altruistic donation was provoked by *le drame du sang contaminé*, in which over two thousand lawsuits were filed by patients who had received transfusions infected with HIV. (It's not all roses in the French *jardin.*)

Two major statutes and a Constitutional revision later, however, altruism in blood donation remains dominant in public policy, as does the concept of solidarity on which it rests. In fact, it's been said that the debate over HIV-infected blood anchored that principle on an even firmer footing in law, establishing that society owed a debt to the victims of technological 'progress'. Lucien Sève believes that the contamination of the altruistically donated public blood supply was triggered by considerations of profit, which dictated the pooling of blood and the collection of serum from prisoners.[64] If Sève is right, then the 'drama of the contaminated blood' may have strengthened the hand of those who oppose all forms of commercial consideration.

The 'French DNA' case occurred well over a decade ago, but CCNE opinions continue to be quite sceptical about commercialisation and genetic patenting. For example, a 2006 opinion[65] on the commercialisation of stem cell science noted that the patenting of human embryonic stem cell lines—the motor of much research interest in the US and UK, and the subject of over two thousand patent requests by 2002—should not and could not be separated from the question of whether human tissue should be commodified.[66]

We've seen that patent law rests on the argument that what is being patented isn't the tissue of any one person, but rather an artefact created in the laboratory. The CCNE committee dismissed this fine distinction at the very start of its lengthy and careful opinion, insisting on asking how a patent can possibly be issued on a stem cell line, if it's derived from part of a human being. In France, the issue about Hwang's patent applications wouldn't just be whether the science worked, or whether the economy would benefit, but also whether it's permissible to patent a derivative of human tissue in the first place.

The status of the embryo isn't the main or only issue in France, either, as it continues to be in the United States. France permits some forms of embryonic stem cell research[67] but prohibits the somatic cell nuclear transfer method, which the UK's Human Fertilisation and Embryology Authority is actively encouraging through its decision to encourage women to 'donate' their eggs. A 2006 CCNE opinion explains this prohibition as motivated partly by fears of sliding down a slippery slope to reproductive cloning, and partly by concerns over the large numbers of human oocytes that would be required, with concomitant risks to women—those risks that the HFEA skated over.[68]

But this document, like the revised laws on bioethics passed in 2004 after lengthy public debate, is definitely less crusading in tone and content than some of the earlier CCNE opinions: less idealistic, or more pragmatic, depending on which side of the English Channel you call home. Explicitly eschewing a 'public good, private bad' attitude,[69] the committee frankly discusses the need to balance the traditionally French principle of non-commodification of the human body against international competitiveness, European Commission emphasis on creating the most flourishing research space in the world, and the demands of French scientists. In Sarkozy's France, after the elections of May 2007, that second set of interests may gain the ascendancy.

Even before the elections, in February 2007, a consultation day for expert witnesses, convened by the Agency for Biomedicine, dropped some heavy hints that liberalisation was on the way.[70] When they're revised in 2009, the bioethics laws of 2004 will almost certainly allow some form of 'therapeutic cloning', if French scientists don't want to be left behind in the continuing jostle to do what Hwang claimed to have done. That means the existing shortage of eggs will worsen, since they'll be needed for both IVF and SCNT. And that in turn means that the principle of non-payment for eggs or any other human tissue may have to be re-examined. The new minister for education, Valérie Pécresse, is also on record as a strong advocate of SCNT[71]—which is now a criminal offence except by special dispensation.

Furthermore, the idealistic stance taken against all forms of commodification of tissue on an official level in France can't prevent individual French men and women from engaging in body shopping outside France. Those Spanish students we met in Chapter One often sell their eggs to French couples, who are among the private Spanish clinics' most frequent customers.[72] Restrictions on who qualifies for IVF in France, together with protective rules limiting egg donation to women who are already the mother of at least one child, have created a boom in reproductive tourism across the Pyrenees.

Despite some slippages and contradictions, however, France, like Tonga, provides proof that resistance to body shopping isn't just futile. Unfortunately, though, the dominant biotechnological economy, the United States, is dominated by policies on genetic patenting which are blatantly designed to favour biotechnology firms or major research institutions,[73] and the pace of developments in their favour is, if anything, getting faster all the time. In the next chapter, we'll see that their interests are also privileged over those of the patients and families who provide the samples, and, increasingly, over those of the researchers who make the discoveries.

6

The biobank that likes to say 'no'

In the *Moore* case, Dr Golde and the University of California were in collaboration—or cahoots—to develop the $3 billion cell line made with John Moore's tissue. But as the stakes in body shopping rise ever higher, alliances between researchers and their universities are splitting apart, in an embarrassingly public fashion. The one thing that seems clear so far is that the people who provided the original tissue don't benefit much.

That includes you and me, of course. Almost all of us have provided tissue to biobanks, whether knowingly or unknowingly. Whereas the *Moore* case concerned one individual with a particularly valuable cell line, tissue and genetic data biobanks comprise contributions from thousands of individuals, not all of them necessarily cell line superstars like Moore.

The odds of any of us having as rare and valuable tissue as Moore are minimal, and so you might think we're protected from the indignities he suffered, as did Henrietta Lacks's family. But you might be wrong. Even biologically run-of-the-mill individuals are likely to have some of their tissue stored in biobanks. That likelihood is enhanced by the sheer size of collections like the new UK Biobank, which will take samples from 500,000 individuals aged between forty and sixty-nine. Furthermore, biobanks created from scratch with the consent of the donors, like UK Biobank, are vastly outnumbered by biobanks formed of existing material, created without explicit consent in many cases. In 1999, a 'conservative estimate' put the number of stored tissue samples in the United States at over 307 million, from more than 178 million people.[1] At that time, the quantity of samples was thought to be increasing at a rate of over twenty million a year. So for Joe and Josephine Public, there's an excellent chance that some of their tissue is lying in a biobank somewhere, even though they know nothing about it.

In the United Kingdom, the Retained Organs Commission, appointed in the wake of the Bristol and Alder Hey hospital scandals concerning tissue stored from dead children without their parents' consent, also uncovered large tissue banks at many other hospitals and academic institutions.[2] Body parts from over 850 infants and children were stored at Alder Hey, on the secret orders of a research pathologist, Dr van Veltzen. More recently, in 2007, it emerged that organs were taken from deceased workers at several UK nuclear power and research sites, including Sellafield in Cumbria, Aldermaston in Berkshire and Harwell in Oxfordshire, and retained for up to thirty years, all without their families' knowledge.[3] At Sellafield, hearts, lungs and other major organs from at least sixty-five employees were stored.[4]

Many of the samples held in tissue biobanks, however, are no more than blocks or slides containing tiny amounts of tissue, and often the sampling involves no additional procedure or risk to the patient. In that case, what kinds of ongoing rights should patients have?—assuming they ever find out that bits of them are hovering under a microscope somewhere. Does it matter if they *don't* find out? As with genetic patenting, the actual physiological harm done might seem minimal, especially if the amount of tissue is minuscule.

Of course, the *Greenberg* case in Chapter Two demonstrated that sometimes it matters a great deal to patients and their families. The Canavan judgment allowed them very few rights over their children's tissue, even when the biobank was created largely through their efforts. There the researcher's employer, Miami Children's Hospital Research Institute, was awarded the meatiest bits, including valuable patent rights, while the Canavan families had to content themselves with the off-cuts and gristle of minimal damages. And whereas the French government intervened to protect families from possible exploitation, the US Federal courts largely took the other side.

Although he was cited as a co-respondent in the *Greenberg* case, together with his then employer, the Miami hospital, Dr Matalon actually claimed that he'd never personally made any money from the patent line and other rights in the Canavan children's tissue. However, that wasn't a major issue in the case. Effectively, the researcher and his hospital acted as one. But what would have happened if they'd fallen out over the spoils? That seems more and more likely, as the stakes rise. What will happen to patients and their families in the scrimmage?

POSSESSION IS TEN-TENTHS OF THE LAW: THE CATALONA CASE

In the more recent case of Dr William Catalona, the researcher-employer alliance ruptured spectacularly into litigation. Catalona, a respected urologist and surgeon who developed the prostate-specific antigen (PSA) test for prostate cancer, created a research tissue biobank, containing material from large numbers of his patients. Based at Washington University in St Louis, Missouri, the GU (Genito-Urinary) Repository grew to encompass 3,500 prostate tissue samples, of which between 2,500 and 3,000 came from patients who had been treated by Catalona during his quarter-century of employment there. Over 270,000 serum, blood and DNA samples were stored in the biobank, which was funded in large part by Catalona's success in obtaining grants.

Although the university administered the grants, it's probably fair to say that it was Catalona's effort and status which brought the money in; that's usually the way with research funding. In fact the university was probably being paid by its employee Catalona, rather than the reverse. Major universities on both sides of the Atlantic now depend on their professors to subsidise them through grant acquisition. When I was employed in the medical school of Imperial College London—rated the second-highest university in the UK at the time, behind Cambridge but ahead of Oxford—we were expected to earn our own salaries back in grant money. I was also teaching large numbers of undergraduates and administering the Medical Ethics Unit, but the university was effectively receiving these services free, courtesy of the body that provided my research grant.

Both Washington University and William Catalona assiduously denied that their motivation was monetary. Catalona was depicted in the *New York Times* as a committed, patient-centred researcher, so innocent of the tawdry world of megabiobucks that he had never even bothered to take out a patent on the PSA test.[5] The university claimed that it, in turn, 'has never profited a nickel from the repository, although it has spent hundreds of thousands of university dollars and federal research dollars to create and maintain the repository and conduct research using it'.[6]

On the whole, Catalona's claims came across as more genuine. The university may not have 'profited a nickel', but only because Dr Catalona was insufficiently business-minded, in its managers' view. The institution's agenda was made obvious in an email from John Kratochvil, Washington University's business development director, to Theodore Cicero, vice-chancellor for research:

> Bill Catalona wants to send nearly 2000 documented samples to Hybritech for free. Just from a cost recovery scenario, this should be worth nearly $100,000 to the University. The only consideration Hybritech is offering is the potential for Catalona to get a publication. It is my opinion that this is an unacceptable proposal.[7]

'Cost recovery' was a specious argument: the only cost to the university was half an hour of a lab technician's time, plus the postage for shipping the samples. All other costs—researchers' salaries and storage costs of the tissue—had already been covered by either Catalona's grants or the patients' insurers.[8]

Catalona was only proposing to do what researchers are traditionally meant to do: share information and materials freely with other scientists and doctors, in the interests of progress. That's what was deemed 'an unacceptable proposal'. Altruistic exchange of expertise and knowledge is no longer sacrosanct, even (or especially) for universities. Scientific collaboration has been toppled by a new idol, 'cost recovery'.

In biotechnology, it's common for public funding to cover the risks and costs, but for private firms and universities to reap the proceeds. Control of large biobanks also means the potential for profit-making patents—as we saw in both the *Moore* and *Greenberg* cases—and for material transfer agreements, licensing use of the database to other institutions. The old days of free transfer of materials between high-minded scientific researchers are gone for good; that principle may have applied when George Gey altruistically made the HeLa cells available round the world, but no longer. And all this rests on the patients being seen as having given their tissue, freely and finally, to the institution running the biobank.

It was also Catalona's reputation in the field of prostate cancer treatment which attracted the patients, and their tissue samples, in the first place. 'Their intent [was] to work with the investigator, not the institution.'[9] As one patient, Richard Ward, said:

> [Washington University was] where Dr Catalona was, so that's where I was. I was looking ... for the best to do my surgery, the best in the world if I could find them and they were available to me. I was willing to wait for him if I could get him.[10]

Another patient, James Ellis, added:

> I have six grandsons and the one thing I want to do is what I can to make certain they don't go through what I've gone through, and my family's gone through, for the last fourteen years. And I [can't] think of anybody that I would have more faith in to do the kind of research that might help my grandsons on my samples, my tissues, my body parts, than Dr Catalona.[11]

But when Catalona changed employer and location, from Washington University in St Louis to Northwestern in Chicago, he found himself unable to take the tissue bank with him for his future research. Letters sent to former patients resulted in six thousand men agreeing to Catalona's request that their tissue samples should move with him to his new job. Yet all their affirmatives were cancelled out, it appeared, by one massive negative from Catalona's employer.

To rephrase the well-known slogan of Lloyds TSB—'the bank that likes to say "yes"'—Washington University's GU Repository is 'the biobank that likes to say "no"'. The university went to court to try to prevent Catalona from moving the tissue collection, and Catalona fought their action. Would the combined weight of the patients' numbers and his expertise pack the necessary legal punch?

The answer from Judge Stephen Limbaugh came loud and clear: absolutely not. Washington University obtained a court order prohibiting the doctor from taking his tissue bank with him, contrary to his patients' express wishes. The tone of the judgment, delivered at the Federal District Court level in April 2006, was that Catalona was little more than a flunkey (and a disloyal one at that), while his patients were just exhibiting blind loyalty.

Possession, that famous nine-tenths of the law, determined the judgment entirely in favour of Washington University, making it more like ten-tenths. Research participants retain no ongoing rights in their donated materials, in the court's view: certainly not financial rights to profit from the licensing, patenting or sale of their samples—which were never at issue—but not even the weaker rights to control where research is performed with their tissue, by whom and on what.

Professor Lori Andrews, one of the best-known US experts on commercialisation of human tissue, called the ruling 'a big setback for patients' rights', and so it would seem to be, at least on first reading.[12] Even with the support of the doctor involved—which neither Moore nor Greenberg enjoyed—patients failed to establish rights to control the 'downstream' use of their tissue. Those rights are increasingly vital to anyone who gives tissue, as the future uses grow ever more complex and profitable. Catalona's patients were trying to marshal some control over what happened next to their tissue, just as Moore was. But while Moore failed, Catalona's patients must have thought they had a good chance of success, because their doctor was on their side.

However, the Catalona story is actually a great deal more complicated than it might appear from these bare facts, as are the issues raised by the case. First, could Catalona be said to have exercised undue influence by writing to ask his patients' permission for their tissue to be transferred to his new employer? Hwang is believed to have asserted illicit pressure over his female researchers to 'donate' eggs for his research, and Golde exerted power over Moore in allowing him to believe that his ongoing treatment depended on supplying the tissue samples that Golde wanted so badly. Did Catalona's actions likewise constitute illicit influence over vulnerable parties? Judge Limbaugh certainly thought so:

> In connection with their argument that the RPs [research participants] 'own' their excised biological materials and can transfer them to any institution or person of their choosing, Dr Catalona advances the notion that his letter and 'Medical Consent and Authorization'

> form allegedly signed by approximately 6000 RPs effectively legally carries out the RPs' 'right to discontinue participation' ... [T]he context in which this form was sent is troubling to the Court. He sent it to RPs, many of whom were his patients and emotionally tied to him, advising them of his move, of his desire to continue his consultation/treatment practice, and then describing his need to use these samples to further his help to them. Such a communication smacks of undue influence.[13]

On the other hand, Judge Limbaugh seemed to be inferring that the patients' trust in Catalona was automatically a bad thing, and that they were too emotional to judge rationally. Although he was criticising Catalona for medical paternalism, you might say that he himself was indulging in paternalism and that his view of the patients was condescending:

> Although the Court respects the testimony given by Messrs. Ward, Ellis and McGurk, it was clear that these gentlemen all had a deep personal connection to Dr Catalona, and believed that they owed their lives to him. The Court understands and appreciates these feelings, but their testimony ... is suspect ...[14]

True, the old-fashioned paternalistic doctor-patient relationship can certainly give rise to abuse, but did it in this case? Are patients automatically debarred from making rational judgments about the ongoing use of their tissue? That seems offensively patronising, and likely to ensure that patients can never succeed in any attempts to control the later uses of their tissue. Actually, if Ward, Ellis and McGurk were so tied up in their emotions about their illness and treatment, ought we to infer that they weren't competent to make a legally binding gift of their tissue in the first place? That wouldn't have suited Washington University's purposes at all.

The donor's intention is crucial in establishing the conditions surrounding a gift, but Judge Limbaugh overrode actual evidence about the donors' motivation, on the grounds that these men were too personally indebted to Catalona for their evidence to be valid. On that logic, they were so desperately irrational as not to be able to make a valid gift at all—in which case the transfer of their property interests to Washington University was presumably also invalid.

Yet the wording that Catalona used in his letters was certainly ambiguous and ambivalent:

> You have entrusted me with your samples, and I have used them for collaborative research that will help in your future medical care and in the care of others for years to come ... [To continue this work] I need your assistance and your permission.

Although the form sent for patients to return clearly stated that the samples would only be used for research purposes, not for any one patient's treatment, it's quite possible that patients were swayed by that line in the letter about 'collaborative research that will help in your future medical care'. Patients are known to be jittery about such things: think of what Moore said about his continued dependence on Golde, and his natural unwillingness to cross his doctor.

What Catalona said was true, though: if the patients' samples were anonymised—as the university intended to do—all links to particular patients would have been lost. The sorts of research which could then be performed would be diminished, since the sample information would be separated from the patients' ongoing medical records. Any opportunity to use research findings to benefit those particular patients would have vanished as well.

Judge Limbaugh also stamped down firmly on the claim that ownership rights of any sort could persist after the patient had given informed consent to the donation of his tissue: 'A completed *inter vivos* gift [a gift between living persons, as opposed to a bequest after death] cannot be revoked once the gift is delivered and accepted by the donee.'[15] And, the judge said, because the consent form was printed on 'Washington University' headed paper, it should have been clear to the donors that the university, rather than the doctor, was the donee.

That's the sort of narrow argument that gives lawyers a bad name. As Lori Andrews remarked: 'If a woman donated her kidney to her brother, after signing an informed consent form on a university's letterhead, she could reasonably expect that the kidney would be given to her brother, not used by the university for whatever purpose it chose.'[16] Clearly, the judge intended to use informed consent as a knock-down argument, to dismiss the patients' claims out of hand.

It's hard to avoid the conclusion that where biobanks are concerned, informed consent is a hollow reed, a fig leaf, or whatever botanical metaphor you prefer. Originally intended to refer to a single intervention by a single physician, informed consent can't plausibly be stretched to fit biobanks' requirement for blanket permission, approving multiple uses by multiple users.[17]

Catalona's patients testified that they had no intention of making an unconditional gift to the university, and Catalona himself denied having received an unconditional gift. The patients reiterated that in fact the consent form said nothing at all about a 'gift', either conditional or unconditional. They denied that they had intended to relinquish all 'dominion and control' over their tissue, as a valid gift requires. Because the party claiming the gift has the burden of proving that intent,[18] it should have been up to Washington University to offer convincing evidence. Instead, the donors had to make all the running.

It seems plain that the judge was reading 'gift' into the situation, in the same way that stem cell researchers or IVF centres prefer to use the language of 'donation'—even when the relationship is really barter, as when women are given cheaper IVF treatment in exchange for 'sharing' their eggs. All these deliberate uses of altruistic language perform two neat tricks. First, they make the *patient* look like the mercenary one, if she tries to assert any property rights in her own tissue. Judge Limbaugh figuratively threw up his hands in horror at the idea of allowing patients to control the disposition of their tissue, because then 'these highly prized biological materials would become nothing more than chattel going to the highest bidder'.

Meanwhile, back in the non-judicial real world, a senior US government scientist was recently found to have trousered $600,000 in 'consulting fees', paid to him by a drug company in exchange for tissue procured at public expense in a $6.4-million government-funded project. Dr Troy Sunderland, chief of the old age psychiatry branch of the National Institute of Mental Health, provided Pfizer with 3,200 tubes of spinal fluid and 388 tubes of plasma for research into Alzheimer's disease: otherwise known as 'chattel going to the highest bidder'.[19]

The second neat sleight of hand, in the Catalona case, was the way in which altruistic language enabled the judge to invoke a legal concept, gifts *inter vivos*, in order to assert the finality of the 'gift' and the patients' absolute lack of rights thereafter. That doctrine was in fact challenged by Catalona's attorneys, who contended that the donor of a charitable gift *does* have the right to attach conditions, and that a departure from those conditions renders the gift forfeit. For example, if a benefactor provides a sum of money to a university on the understanding that a building will be dedicated in his name, but the university siphons the funds into a general building maintenance account, the donor is entitled to have his money returned.[20]

The donor's intention is crucial in establishing the conditions surrounding a gift. Catalona's attorneys argued that the research participants had only permitted the use of their tissue in prostate cancer research conducted by Catalona. They had not intended to transfer all rights of ownership over their tissue. But Judge Limbaugh overrode this actual evidence about the patients' motivations, on the grounds that these men were too personally indebted to Catalona for their intentions to be valid. On that logic, as we've seen, they were really so desperately irrational as not to be able to make a valid gift at all—in which case the transfer of property interests to Washington University was presumably also invalid.

TWO STEPS BACK OR ONE STEP FORWARD?

Although the initial judgment in the *Catalona* case went against both the surgeon and his patients—which you might see as two steps back—in an important sense the case represents progress from both the *Moore* and *Greenberg* instances: one step forward. Unlike those earlier instances, the judgment in *Catalona* started from the premise that separated tissue *can* be the rightful object of property rights. The question is whether giving that tissue away has the effect of extinguishing those rights.

The judge did at least recognise that tissue, once separated from the body, now needed to be seen as something in which property rights could be vested.[21] This conclusion was really forced upon him by the facts of the case: unlike *Moore* and *Greenberg*, which concerned patients' claims to patent rights and profits derived from their tissue, *Catalona* was about rights in the tissue samples themselves.

By 2006, when the first judgment in *Catalona* was handed down, it had finally become impossible to maintain the old common law position that tissue taken from the body is either abandoned or 'no one's thing'. Although Judge Limbaugh held that it was most definitely someone's thing—the university's—he also dispelled the old fiction that excised tissue can't be seen as valuable property. Once tissue is recognised as having value, and as being a kind of property, we can deal with it in other ways than the increasingly false gift metaphor—and a number of new models of property in tissue are being devised to do just that.

Many judges, like Limbaugh or the *Moore* majority, distrust the idea that patients can own their tissue or retain any rights of control after 'giving' it away, because they wrongly perceive property as an all-or-nothing concept, although the common law actually considers property to be a 'bundle' of distinct rights. If Moore benefited from rights to income, capital value and sale proceeds of his cell line, these judges opined, there would be no incentive for research sponsors or firms to develop the cell line for their own commercial purposes, as well as for the benefit of society. Further, it was felt inequitable to allow Moore to enjoy income or capital value from his T-cells, when it was only by good fortune that he happened to possess a particularly effective immune system.

But money, as they say, isn't everything. It wasn't Moore's prime motivator. And what Catalona (along with his patients) most desired was the right to physical possession of the tissue, the right to determine how other researchers used it and the right to be protected against someone else taking it. Some of the consent forms that the patients had signed stipulated that they wouldn't receive monetary compensation, but none of the forms barred them from claiming these other rights. By default, you might think, they retained them, since they hadn't explicitly transferred them.[22]

At no point had Catalona tried to make a profit from the prostate tissue itself, or from a patent on the PSA test, which he could easily have done. So he probably would have been perfectly willing to allow Washington University the rights to income, capital value and sale proceeds. The university's management, in turn, having assiduously insisted that they never 'profited a nickel' from the GU Repository, would no doubt have been happy with those rights to income, capital value and sale proceeds, but less concerned about powers to determine how others used the biobank in their research—provided the university's need for 'cost recovery' was respected (at a genuine rather than a trumped-up level). Catalona himself could then have continued to use the biobank for his further studies, as his patients wished. (In the poisonous aftermath of the lawsuit, Catalona is probably the last person Washington University would now allow to use his own biobank.)

It can't be beyond the considerable ingenuity of the biotechnology sector to devise legal agreements that allow one party to keep physical possession and the right to management, while allowing a second party to make the profits. Weirder and more wonderful licensing, patent and use agreements are devised every day. But as long as judges like Limbaugh interpret such cases as winner-takes-all, there's no incentive to do so.

Winner-takes-all is a loser of an argument. A better model for such an agreement is a charitable trust. The notion of a trust, already used to protect minors and others who can't defend their own interests, could go some way towards plugging the rights vacuum for biobank patients. Even if contributors to biobanks don't possess full proprietary rights, as do the beneficiaries of a trust, the trust model stresses the duties of administrators of the biobank, while severely restricting their own property rights.

The notion of the trust as a model for biological repositories was first mooted by Karen Gottlieb in 1998[23] and further developed in an influential 2003 article in the *New England Journal of Medicine* by David and Richard Winickoff.[24] The charitable biotrust sets out a far more precise programme of duties and entitlements than the rather vague notions of 'stewardship' and 'custodianship', used by many biobanks which are actually more like brokers to the private sector.[25] David Winickoff is now working with the Veterans' Administration, the largest provider of publicly funded medical care in the United States, to apply aspects of the charitable trust model to genomic databanking.

Under a trust agreement, the donor (or settlor) formally transfers her property interest in tissue to the trust, appointing trustees who have legal duties to use the property for the benefit of a third party (the beneficiary). In charitable trusts, the beneficiary has to be a *class* of persons (neither an individual nor the community at large). Such a collective grouping might be as broad as all of a country's health service patients, or as narrow as sufferers from a particular disease.

Each donor sets up an individual trust instrument, assigning certain property interests to the same trustee, a non-profit organisation that holds and manages the biobank in accordance with the agreed charitable purpose. Full disclosure of all pending commercial arrangements must be made to the settlor at the time she gives her agreement. If the biobank fails or goes bankrupt—a real risk in the easy-come-easy-go world of modern biotechnology[26]—then its assets can't simply be transferred to the highest bidder or a creditor.

Unlike corporate executives' legal obligations to their shareholders to maximise profits, the fiduciary duties of trustees aren't primarily profit-oriented. Donors can be protected from unwanted commercialisation of their donations, or from transfer without any secondary consent to an unknown third party. They may also appoint representatives from their number to the board of trustees, which mitigates the paternalistic nature of the trust.[27]

In other words, the people who gave the tissue get to work with the people managing its use. They have a measure of control—which is almost certainly what they really want. At least, so it seemed in the very diverse cases of Henrietta Lacks, John Moore, the Canavan parents and the Tongan nation, as well as in the Catalona example.

It's been argued that only giving patients property rights in their own stored tissue will force researchers and institutions to treat them less arbitrarily. As the British legal academic Roger Brownsword remarks: 'Whereas the voices of those who have property rights simply cannot be ignored, the voices of those who have an interest, but without the backing of property, are just so much "noise"'.[28] The charitable trust model gives some hope that the voices of patients who give tissue to biobanks could be heard *together* with those of universities and biotechnology companies, rather than drowned out by them.

CATALONA REVISITED: THE APPEAL COURT JUDGMENT

For all its faults, the first Catalona judgment *did* recognise that patients had some rights in their tissue, if only sufficient to give that tissue away. But would that view be upheld if the case went further?

Go further it did, when in July 2006, three months after the district court judgment, a coalition of interested parties filed an appeal against Judge Limbaugh's holding. The appellants included Catalona himself, eight of his patients and (as *amicus curiae*, a 'friend of the court' offering an advisory expert opinion) the People's Medical Society, the largest patients' advocacy group in the United States. That organisation had a long history of involvement in 'body shopping' cases, dating back to *Moore*, when it succeeded at least in persuading the court that Moore had a claim for lack of informed consent and breach of fiduciary duty. As the *amicus curiae* brief put it:

> Medical institutions and physicians have a strong pecuniary and academic interest in asserting an unprecedented claim of ownership of tissue used in research because they can sell the tissue to other researchers and can profit from any new discoveries. In contrast, the People's Medical Society has no such pecuniary interest and thus can raise fundamental legal and policy considerations that will not otherwise be addressed.[29]

Limbaugh had been culpably naïve, the People's Medical Society implied: 'The district court misunderstood how the research enterprise operates and the policies that undergird that system.'[30] Research institutions are no longer disinterested servants of science, but major corporate 'players', even if formally not-for-profit. Washington University claimed net assets of over five billion dollars in 2003. Its highest-paid physician/researcher—not Catalona himself—received over one million dollars in salary that year.[31]

Yet from the testimony about who bore which costs in the *Catalona* case, Washington University is the Tommy Hilfiger of the biotechnology sector, adding little to the value of a product except its own 'brand'. (The Hilfiger firm outsources production abroad and does nothing except sew in its own labels, which doubles the sale price of the product.[32]) There may be some instances where the employer or research institution bears risks and costs, but this wasn't one of them. So the winner who took all in the initial *Catalona* judgment, Washington University, actually had the least moral right to win, on the grounds of having put work in.

Once patients realise that the legal system doesn't protect them, but in fact reinforces the already considerable might of the big institutions, they will refuse to donate their tissue, and research will judder to a halt. As the *amicus curiae* brief argued, pragmatic reasons actually favour granting more substantial property rights to participants. This isn't 'just idealism': it's sound political sense:

> This legal scheme [protecting patients' rights] not only makes good moral sense, it also makes good practical sense. Most people are willing to participate in research when they can choose the type of research in which their tissue is used and can stop participating if they no longer desire to ... We would not have had a breakthrough in AIDS treatment if AIDS patients had not been able to choose the type of research that was done on them and their tissue. We would not have learned about sickle-cell anemia if African-Americans had not been allowed to choose to provide their blood for research on that particular disease.[33]

Without patients' trust in the research system, however, there can be no further progress, as the brief went on to claim:

> If the district court's decision is upheld, the precedent will not only disrupt medical research, but other important transactions because:
>
> 1. Patients and research patients can no longer rely on the promises made by physicians and research institutions in informed consent documents ...
> 2. Research participants will lose the right to stop research on their tissue when a research institution materially changes the nature of the research being performed ...
> 3. Institutions with biorepositories will be free to sell tissue to the highest bidder, conduct any type of research on the tissue and generally treat the tissue as their own property, free of the rights of the person from whose body the tissue was obtained.[34]

When patients have no control over 'downstream' uses of their tissue, when they observe that tissue which they have donated altruistically becomes 'big bucks', then cynicism will be the likely response. For that reason—again out of pragmatism rather than idealism—a committee of the US National Academy of Sciences warned that researchers and their universities shouldn't seek blanket consent from patients: 'It is not ethically or legally acceptable to "consent" to future yet unknown uses of their identifiable DNA samples.'[35] As the appellant brief remarked, this is the logical conclusion, if what participants have granted is only a specific *use* of their tissue, not full-blooded ownership rights for the recipient.

Concluding its arguments, the *amicus curiae* brief noted:

> This case began as a simple employment dispute between a researcher and a research institution, but it has expanded into a case where the future of medical research in this country is at stake ... It is even possible that people will not seek out medical care if they cannot trust that their tissue will not be taken and used in medical research against their will.[36]

That claim was something more than just legal rhetoric. The authors could actually have gone further: the validity of the 'no property in the body' principle was also at issue, and with it, the future of body shopping. From one American case in the apparently arcane area of biobanking, we might have derived a radically new answer to the question: 'What makes you think you own your body?'

We might have, but we didn't. On 20 June 2007, three months later than the judgment was expected, the United States Court of Appeals for the Eighth Circuit issued its opinion, affirming what they called the 'well-reasoned opinion and judgment of the district court'.[37] No doubt this verdict was a matter of considerable relief not only to Washington University, but to all the other universities and medical research institutions which had also filed *amicus* briefs in support of the college: the American Cancer Society, the Mayo Clinic, the American Council on Education, and Cornell, Duke, Emory, George Washington, Johns Hopkins, Stanford, Michigan, Minnesota, Pittsburgh and Rochester Universities, as well as the Association of American Medical Colleges and the Association of American Universities. Clearly, major interests are at stake.

For patients, their advocates and others who had hoped for some recognition of patient control over further uses of tissue, the only morsel of comfort lay in the narrow construction which the court circumspectly chose to place on the question of:

> ... whether individuals who make an informed decision to contribute their biological materials voluntarily to a particular research institution for the purpose of medical research retain an ownership

> interest allowing the individuals to direct or authorize the transfer of such materials to a third party. Under the facts of this case, the answer is 'no'.[38]

Under the facts of a more favourable case, might the answer in future ever be yes?

7

Buying the 'real me': shopping for a face

'Just a chin-tuck': that was meant to be the latest in a series of cosmetic surgery operations for the novelist Olivia Goldsmith. Latest and, as it turned out, last: Goldsmith died on 15 January 2004, at the age of fifty-four, from complications following anaesthesia.

As the author of the best-selling novel and film *The First Wives Club*, Goldsmith had dissected—if that's not too ironic a term—the clinical coldness with which some men cast off their ageing first wives. In fiction, those women took their revenge in exhilaratingly comic fashion; in reality, Goldsmith willingly made herself into a victim of the potentially fatal quest for youth's elixir. As a neighbour related, 'She was the first person who ever talked to me about Botox, and she had it long before it became *de rigueur*.'[1] Against the accusation that Goldsmith was unusually obsessed with looking younger, her friend, the television news anchor Kelly Lange, fired off the comment: 'Who the hell wants to look old? I don't think Olivia was any different from any of us.'[2]

Who can resist the temptation of shopping for a new face? Not, it seems, the *New York Times* writer Alex Kuczynski, despite her perceptive analysis of the rise and rise of cosmetic surgery:

> We have begun to think of our bodies as something like an accessory that can be modified when necessary, discarded when it is worn out, and upgraded when required, a leathery sack to transport us from one medical specialist to the next.[3]

Yet Kuczynski herself came fearfully close to becoming a 'beauty junkie'. Finally pulling back from her adventures with liposuction and eyelid tucks after a disastrous lip-plumping procedure, she wrote: 'The fundamental paradox of this kind of addiction, if one can call it that, is that I was involved in a process of self-improvement.'[4]

Still in her thirties, Kuczynski realised that rather than finding her 'improved' self, she had actually lost her true personality:

> The qualities I had cherished—fortitude, endurance, practical Yankee good sense—had withered, while I had grown as vain and silly as any teenage pop star with lip implants and a bad boob job. Worse still, my very perception had become warped.[5]

But in tune with Kuczynski's own belief that she herself was motivated by 'self-improvement', most of the women she interviewed continued to insist that they were only looking for their truer, better selves, under the crow's feet and naso-labial lines. As one woman exclaimed after her 'extreme makeover' operation: 'Oh my God, I finally look like me!'[6]

Nor is the odyssey in search of perfection undertaken only by Americans. Like other forms of body shopping, such as the IVF industry, cosmetic surgery operates on a global scale. A South African company called Surgeon and Safari offers cut-price liposuction or tummy tucks followed by a week's safari (plus painkillers), much as hospitals in India and elsewhere offer Western patients in need of organ transplants a package of transplant plus post-operative R and R in Goa. Other countries which have developed cosmetic surgery plus holiday packages include Honduras, Jamaica, Brazil and Malaysia. Ireland, with its kinship links to the United States, has been particularly targeted by American plastic surgeons.

Spain, familiar for its glut of for-profit IVF clinics, also has the highest *per capita* number of cosmetic surgeons in Europe. The Costa del Sol welcomes Europeans of all nationalities to its myriad 'rejuvenation centres', such as the Molding Center, where the Nigerian first lady, Stella Obasanjo, died abruptly at the age of fifty-nine after undergoing liposuction.[7]

Fallout from negligently performed cosmetic surgery on medical tourists is fast becoming a problem for their home health care services, which have to pick up the pieces. In Australia alone, a recent survey of sixty-eight plastic surgeons revealed one hundred cases of botched operations from cosmetic surgery package holidays in Malaysia and Thailand.[8]

Despite the risks, cosmetic surgery is becoming the most popular and obvious form of body shopping: buying a new face—one that will convey not a new identity, but the 'real me'. Just how far will the quest for 'the real me' go? Media coverage of two recent face transplants in France may have left many readers thinking that it is only a matter of time before we can all go 'face shopping', with the same insouciance we show in buying cosmetics. Are face transplants likely to be the next area of business interest in biotechnology? If so, what's wrong with that? If the recipient consents and the technology is available, what ethical objections can there possibly be? We'll return to those questions later in this chapter.

'VENUS ENVY'[9]

Bioethicists tend to dismiss the 'yuck factor' as an irrational response to new developments in medical technology, but it's unavoidable in cosmetic surgery. The collagen used to plump Kuczynski's lips probably came, like a great deal of collagen used in such procedures, from the cells of a baby boy's foreskin—now developed into a stem cell line owned by a Californian firm, Inamed Aesthetics. 'Cosmoplast', a synthetic collagen, is made from the cells of a boy born in the early 1990s.[10] (The young man, as he now is, remains happily unaware that his genital tissue can be found in the lips of women around the world.) Cadavers are another source of collagen, harkening back to the scandal around Alastair Cooke's bones.

Breast augmentation surgery is routinely given to American high school girls as a graduation present from their parents. If they realised that the chances are those young women will require regular operations for the rest of their lives, to deal with the after-effects, those loving parents might prefer to revert to the more traditional gift of a second-hand car. It's been estimated that a woman who has breast 'enhancement' surgery at eighteen will need to have surgery twice a decade for the rest of her life, because of the problems of leakage, infection and formation of scar tissue.

Ever inventive in seeking out new markets and undreamt-of procedures, the cosmetic surgery industry in the United States alone was worth between $13 and $15 billion in 2005.[11] Between 2003 and 2004, there was a 44 per cent surge in cosmetic surgery procedures or, as the industry and some academics prefer to euphemise it, 'aesthetic surgery'.[12] Demand in the United States is predicted to grow by 11.2 per cent annually, with one-stop cosmetic surgery parlours operating in shopping malls and advertising in the Yellow Pages.

A specialist lender, the Capital One cosmetic surgery lending site, offers loans tailor-made for the purpose, enabling the patient to buy the face of her dreams now: why wait? Even impecunious college students can qualify for these 24 per cent rate-of-interest loans—which they can always pay back by selling their eggs. Body shopping has created a closed-loop economy of its own.

New 'products' in cosmetic surgery include umbilicoplasty (navel 'enhancement'), breast nipple enlargement, and toe shortening to fit strappy Jimmy Choo sandals. Beauty magazines tout the procedure of 'labial rejuvenation', in which the lips surrounding the vagina are snipped into pre-childbirth form. Or perhaps the 'enhanced' body itself should be seen as the product, and the surgeon's art as a particular 'brand'. Kuczynski tells the tale of meeting a colleague who had recently undergone breast augmentation surgery at the hands of a chic New York surgeon, Dr David Hidalgo. Just as wealthy women call their expensive Manolo Blahnik shoes 'Manolos', this woman flaunted her surgically enlarged breasts, which she proudly called her 'Hidalgos'.[13]

'Aging faces, flat breasts and small penises, which as facts of life were considered unworthy of medical attention, have been progressively redefined as problems worthy of medical consideration and more recently as pathologies or deformities requiring medical solutions'.[14] This medicalisation of perfectly normal features turns them into something 'other' than the self:

> The you who feels ugly is linked to the defective piece but is also imaginatively separable. Partly, this double effect of your body that is both 'you' and replaceable feels like a split right down the center of your identity.[15]

Paradoxically, the women—and it is mainly women—who submit to the surgeon's knife typically claim that they've found a new identity or, perhaps more accurately, have finally found their own original self. The fissure between imperfect reality and perfect self-image has been healed—at least, until the cosmetic surgery industry decides what it will next diagnose as abnormal.

Yet cosmetic surgery, unlike other forms of medicine, is even more predominantly about self-diagnosis. Why do even gender-aware women like Goldsmith decide that there's something so wrong with them that they simply must have surgery? Finding and enhancing 'the real me' is crucial in a society where marketing oneself is seen as inescapable:

> I'd rather spend my money on Botox and a procedure here and there than something that is not a part of me. All we have in life is ourselves, and what we can put out there every day for the world to see. The world is not going to see my great record collection or the stuff I have at home. They're going to see me. And Me is all I got.[16]

The irony is that this glorification of the unique 'I' results in one homogeneous ideal of look-alike beauty. Human cloning is still technologically impossible and universally illegal, but American plastic surgeons are doing their best to remedy that by creating a race of Pamela Anderson clones. Thanks to the mainstreaming of pornography and the promotion of a uniform image of beauty by the media, the successful white female products of plastic surgery all look alike—in fact, very much like the type of woman who can sell her eggs for a premium: tall, blonde, slim but large-breasted. As in Aldous Huxley's *Brave New World*, or in David Mitchell's *Cloud Atlas*, so the dystopian Hollywood world of trophy wives 'is inhabited by identical people who remain young forever because they are replaced with identical copies as they age'.[17]

If any woman who doesn't look like that ideal is defined as abnormal, there's a potentially huge market—particularly if her psyche can be made to seem as much in need of repair as her body. '[T]hat implies from the start that the psyche or the body is somehow broken or disfigured and must be fixed if the actualized, "real" human being is to emerge.'[18] As the writer Carolyn Latteier said of her 'abnormal' lack of a full bust: 'My discomfort with small breasts was more than cosmetic. I felt the lack as a poverty of being, as if my very nature were somehow stark and bony.'[19]

'In the world of rejuvenation everything is "real" or at least seems real.'[20] At the same time that new surgeries proclaim the body to be indefinitely plastic, infinitely capable of makeover, women undertake the risks of cosmetic surgery because they feel it will give them back their 'true' identity. There's a powerful contradiction here between the idea that there are no fixed identities—not even those conferred by mortality—and the compelling motivation to find your authentic self.

'IT MAY BE SOMEONE ELSE'S FACE, BUT WHEN I LOOK IN THE MIRROR I SEE ME'

In November 2005, a Frenchwoman, later identified as thirty-nine-year-old Isabelle Dinoire, became the recipient of the world's first face transplant. After being savaged by a dog, Dinoire underwent a partial transplant, using tissue from a brain-dead donor, at the hands of an international team headed by Professor Jean-Michel Dubernard. Interviewed a year later, she remarked, 'It may be someone else's face, but when I look in the mirror I see me.'

Dinoire's case, and a second face transplant performed in January 2007 on a twenty-nine-year-old Frenchman with a severely disfiguring disease,[21] throw questions about psychological and physiological identity into sharp relief. Despite Dinoire's claim that 'I see me', the transplanted face does retain some features of the dead donor; it is a hybrid, not Dinoire's lost 'real' face. The underlying bone structure and remaining musculature of the patient's own face influence the final combination most strongly, but there are also important physiological aspects of the donor, particularly skin and subcutaneous tissue.

The second patient—deliberately unnamed by the surgeon, who was concerned about patient confidentiality—took a rather different view of his new facial identity: 'My brain has taken on two faces.' ('Le cerveau s'est approprié de deux visages.'[22]) This young man thought of both 'before' and 'after' as his faces, whereas Dinoire identified 'me' exclusively with 'after'. Perhaps that's because she had suffered a sudden traumatic injury, a dog bite, whereas his chronic, progressively worsening condition had accustomed him to constantly comparing yesterday's 'before' and today's 'after' in the mirror each morning.

This condition was a genetic malady, von Recklinghausen's disease: a kind of neurofibromatosis, which makes tumours grow on nerve tissue throughout the body. Sometimes these tumours turn malignant, but even if they don't, they can spread quickly. In this case, the entire lower face was affected, to the extent that the weight of the lesions on his lips almost prevented the patient from eating. The young man had already undergone a *quarantaine* of operations, some forty over a ten-year period.[23] Because new tumours cannot grow on the transplanted tissue, according to the chief surgeon Laurent Lantieri, the balance of risks and benefits in this case made a face transplant a reasonable medical bet, although still a tremendous risk. The fifteen-hour procedure began with the removal of the neurofibromas, whose tendency to bleed posed real problems for the surgical team.

Clearly, neither of these cases can be likened to Olivia Goldsmith's 'just a chin-tuck'. Yet a face graft, some commentators think, is just the logical extreme of shopping for the face of your dreams. Face banks might supposedly allow us to shop for the face of a model or a football star. 'Could face transplants become the latest symbol of affluence, the "fashion label" of the early twenty-first century?'[24]

> Commodification of the human body is not new, but facial transplantation may be the most striking example of the results of this reductionist view of the human body. It seeks to resolve the problems of disfigurement by disposing of the damaged face and replacing it with the face of another.[25]

Extreme indeed, but logical? The risks inherent in both partial and full face transplants are far too grave for anyone to undertake the procedure lightly, let alone readily persuade the crack surgical team needed to perform the extremely complex operation. The after-effects of the procedure can be life-threatening. Even with heavy doses of immunosuppressant drugs—which carry their own risks—both of the French patients experienced severe bouts of tissue rejection.

In a partial face transplant such as these two patients underwent, a variety of tissues are involved—so that rather than a single-organ transplant like a kidney graft, they had a 'composite allograft transplant' (CAT). Muscles, mucosa, sensory and motor nerves must all be 'degloved' from the deceased donor and transplanted to the living recipient. In a full face transplant, ears, nose and eyelids might also be included. The procedure is complex, and the benefits limited. Although new skin and subcutaneous tissue may be more flexible than the pre-transplant face, if the recipient is a victim of burns or other trauma, it's unlikely that all the functions can be restored.

And if the face transplant goes wrong—if the body rejects the foreign tissue—it might have to be removed, unless heavier doses of immunosuppressive drugs can overcome the problem. The chances of rejection are high, since the skin is extremely 'antigenic' (that is, reactive against transplanted tissue). In fact, the skin is one of the most antigenic (or 'immunogenic') of all the tissues or organs that can be transplanted. Whereas kidney transplants—much less reactive—have a three-year survival rate of about 83 per cent,[26] it's thought that the immediate chance of short-term rejection for face transplants is about 10 per cent, and the longer-term probability of chronic rejection between 30 and 50 per cent.[27]

That's very high indeed. And if a kidney fails, the recipient would probably be able to go back to the alternative of dialysis. If a face transplant fails, the only choice is removing the entire face (with extremely dire consequences if some underlying tissue has been removed to make the graft 'stick'). In that case, the patient would be left much more disfigured than before. Additionally, a repeat transplant—if a donor could be found—would be even more likely to be rejected by the body's immune mechanisms, because the underlying tissue would have become even more sensitive.[28] Yet the risks of failure are rarely confronted head-on, even in some of the most influential articles: most worryingly, in a much-cited piece by the team at a clinic in Louisville which proposes to do the first face transplants in the USA.[29]

In her transplant, although certainly not in the dreadful injury that occasioned it, Isabelle Dinoire was comparatively lucky: she received an unusually well-matched graft. Out of six possible tissue matches, five were fully compatible—fortunate indeed, given the haste with which a donor had to be found. But she still suffered several major episodes of rejection, the first around the eighteenth day after the transplant. The same occurred with the second patient, whose surgeons were only able to find a donor who matched the recipient on three out of six factors. He also experienced rejection, which was overcome, as with Dinoire, by increasing the dosage of immunosuppressive drugs. And these were episodes of short-term rejection, where the chances of overcoming the immune reaction are better than in chronic long-term rejection, the major cause of overall failure for transplanted organs.[30]

A lifelong regime of immunosuppressive drugs, particularly in continually increasing dosages, carries such high risks that face transplants can actually shorten life. These powerful medications are implicated in a number of potentially serious complications. 'The healthy person with a facial disfigurement is then transformed into a morbidly ill individual who must endure a toxic regime of drugs for the remainder of their life.'[31]

Raising the dosage of immunosuppressive drugs lessens the risk of rejection but increases other hazards. All such drugs carry risks, such as hypertension, kidney damage, diabetes, infection and the long-term possibility of cancers (up to a four-fold increase in the incidence of colon and lung cancers).[32] 'The weakening of the body's defenses allows malignant cells that would normally be destroyed in an immunocompetent individual to survive and multiply in an immunosuppressed host.'[33]

It's hard to believe that anyone would take such potentially fatal risks lightly, particularly a young person. The average life of a transplant is limited, particularly for composite tissue grafts like facial transplants, and the chances of developing rejection accumulate with the years. But in the case of the second French transplant, the patient was already confronting a deadly calculus. His neurofibromas had grown most quickly during his childhood and adolescence. Now, as he neared his thirties, they threatened to degenerate, possibly becoming cancerous.

Although face transplants are arguably so experimental as to constitute research rather than therapy, the Helsinki Declaration, which governs research ethics, allows for more latitude in comparing risk with benefit if the only alternative to the experimental treatment is death or more severe disability.[34] But does this exemption apply to face transplants? That depends on how severely we rate damage to the face. In the extreme, you could argue that massive facial deformity entails a sort of social death. Isabelle Dinoire's first words in an interview given to *Le Monde* after her operation were: 'I've returned to the planet of humankind.'[35]

THE FACE: JUST ANOTHER PART OF THE BODY?

But what's so special about the face? Even though the procedure of transplanting all the multiple components that make it up may be more complex than a single-organ transplant, isn't the face essentially an organ like any other? And if parts of the body like eggs and kidneys are readily sold on global markets, as we saw in Chapter One, why should we be shocked if faces also become commodified? Isn't that just the logic of body shopping?

On that reasoning, the difference between cosmetic surgery and face transplants is only a matter of degree and medical know-how. After all, facial regeneration through surgery worked wonders during World Wars I and II, and later for the 'Hiroshima maidens', who were brought to the United States after the first atomic bomb blast to reconstruct their features. Jaw transplants have been possible since an operation by an Italian surgeon in 2003, but they haven't attracted the same degree of attention as partial face transplants.[36] Presumably full face transplants would be even more newsworthy, but isn't the difference just one of degree?

Does it really matter, either, if we can't draw a firm line between 'necessary' reconstruction and 'mere' enhancement? Is it paternalistic even to try? Isn't it up to the patient to decide? And if the patient can afford it, what would actually be wrong with paying for a new face?—as well as the lifelong immunosuppressants required to prevent rejection, if those aren't covered by a national health service.[37]

Of course the face is a particularly sensitive area, both physiologically and psychologically. If the face *is* the person, as is conventionally supposed, perhaps we're really talking about the dead donor being given a new body, rather than the living recipient getting a new face.[38] In that case, the recipient could even be seen as consenting to the death of her old identity and to the resurrection of the dead donor.

If that seems far-fetched, consider the reaction of recipients of other organs, who commonly report being troubled by a sense that they're no longer themselves.[39] One adolescent girl even refused a heart transplant on those grounds, although her refusal was eventually overborne. She had said: 'I would feel different with someone else's heart; that's a good enough reason not to have a heart transplant, even if it saved my life.'[40]

Criminals were once punished by having their noses removed, to brand them irretrievably as social outcasts. 'Facial mutilation ... was plainly a form of social torture, designed to imprison the offender in a face that instantly signaled the person's sins.'[41] The recipient of the second French transplant felt very strongly that his facial deformity had made him, too, into a social pariah, unable to hold down a job or enjoy a glass of wine in a bistro.

In other cases, however, people with serious burns or other facial 'abnormalities' reject the idea of surgery to 'cure' their condition. While Changing Faces, a British advocacy group for people with severe facial injuries, accepted the French patients' right to choose, they (and other disability rights organisations) have asked whether the mote is in the eye of the beholder. Is our beauty-obsessed society medicalising the 'problem' of 'abnormal' faces—just as cosmetic surgery has turned small breasts and penises into dreadful deformities?

If society were more tolerant of facial 'deformities' which aren't life-threatening, would people like Dinoire still feel such an urgent need for surgery? And if face transplants are offered freely, 'will surgical correction become the expectation?'[42] Will people with facial disfigurement be subject to even more intense ostracism? Will surgeons be tempted to correct whatever is correctable, merely because it *is* correctable?

Cosmetic surgery has already been used to 'correct' the faces of people with Down's syndrome, so these fears aren't just fanciful.[43] After all, the inventiveness of the cosmetic surgery industry knows few bounds, as the example of vaginal labial 'improvement' demonstrates. Like most other industries under late capitalism, it's learned how to create demand out of nothing.

The UK's Royal College of Surgeons has pointed out a cruel irony: 'Those patients who may be best placed to give valid informed consent for the procedure may also be best able over the longer term to adapt to their existing appearance.'[44] Conversely, anxiety about facial disfigurement of a non-life-threatening kind can so undermine some patients' autonomy as to cast doubts on whether their choice of surgery is really free:

> The more vulnerable patients are made by such distress, the more likely it is that this vulnerability will create psychological difficulties in obtaining valid informed consent from them ... This issue is especially relevant to facial transplantation because surgeons involved will inevitably be enthusiasts—rightly so—and patients will know that they are utterly dependent upon them to obtain the transplant that they desire ... Given the desperation that led to the desire for facial transplantation—of accepting, for example, the inevitability of being made unwell by and dependent upon relatively dangerous immunosuppressive drugs for a lifetime—transplant failure could be devastating for even the most psychologically resilient patient.[45]

It's also vital to remember that face transplants aren't just about what the patient wants, or what the surgeon proposes. Because the face is the expression of the person, and because persons have relationships, the interests of those in relationships with both donor and recipient also matter a great deal.

This point has troubled me since I was asked to help formulate a London hospital ethics committee's response to an earlier request, for the first human hand transplant. As I wrote with my co-author, the hand, too, is an expressive 'instrument' of intimacy, skill and interpersonal relationship.[46] For the donor's family, it's deeply unsettling to think that the hand which once touched you may now stroke another body.

That same caution about remembering the relatives applies to face transplants. If you're a relative or friend of the donor, you might have to confront the possibility of one day running into a hybrid of the dead person's face and the recipient's features, walking down the street towards you. Even if that gruesome possibility never transpires, the face isn't just a part of the body like any other—at least, not an internal part of the body. In most organ donations the donor remains anonymous. However, because of the external nature of the face and the large amount of publicity that will surround this surgery, it will be difficult to conceal the donor's identity from family and friends.[47]

The donor may be dead, but that's not the end of the matter, although most working parties and commentators on face transplants have acted as if it were. Yet:

> this is not like the decision to donate other organs after death. And this is because ... the donor will die and in a disembodied form live on. What was the stuff of drama to earlier generations (Hamlet's father's ghost reappearing or the Commendatore returning in Mozart's *Don Giovanni*) is set for contemporary realisation.[48]

For the recipient's family, there are equally troubling questions: for example, should they welcome attempts at contact from the donor's family? Such associations have become commonplace in the United States between families of solid organ donors and recipients; they can play a practical role in bereavement, giving mourners a sense that the dead person has helped a concrete someone else to live. But face transplants don't normally save lives, as transplanted kidneys do.

Nor would every recipient, or their families, want to scan the faces of the donor's kin for traces of resemblance to the new hybrid face. On the other hand, donors' families may prefer anonymity, but recipients and their loved ones may want to know more about where the unaccustomed features came from. Any analysis of the rights and wrongs of face transplants that omits these real questions about previously unimagined kinds of relationships is remorselessly superficial.

A CAUTIONARY TALE: THE AFTERMATH OF THE FIRST HUMAN HAND TRANSPLANT

In an extreme but instructive form, Clint Hallam, the recipient of the first human hand transplant, was also driven by a compulsion to find his 'real me', embodied in his 'missing' hand. That hand was both physically and psychologically, literally and figuratively, missing. It, and Hallam's sense of his own identity, were waiting out there to be found—or so he felt.

It's these troublesome questions of identity and rationality that bedevil all the elective, free-choice surgeries this chapter considers. Is it enough just to say that if people want them, they should be allowed to choose freely? That was the argument made in the Hallam case, but with disastrous consequences.

Hand transplants have now been feasible for roughly ten years: as of late 2006, some twenty-four hand and forearm grafts had been performed, on eighteen patients.[49] (Interestingly, all of the recipients so far have been men.[50]) A survey of the first fourteen transplants concluded that: 'active range of motion of the digits has been surprisingly better than would have been expected based on previous results of replantation, but return of sensibility has been less than optimal'.[51]

However, it has been calculated that long-term side effects of immunosuppression mean that hand transplant patients have an 80 per cent higher chance than normal of contracting an infection, a 20 per cent higher risk of developing diabetes and a 4 to 18 per cent higher risk of cancer (particularly skin cancers, although no cancers have yet been reported in any of the patients).[52] Even with increased dosages of immunosuppressives, the frequency and timing of episodes of acute rejection predict chronic dysfunction and failure of a complicated, multi-tissue transplant, such as grafts of the face or the hand.[53] Solid single organs like kidneys are more tolerant of rejection episodes than composite allograft transplants, but even a single-organ transplant doesn't last forever.

Obviously composite tissue transplantation is vastly more complex and difficult than most cosmetic surgery. Nor is it the sort of thing you might squeeze in between a funeral and the after-funeral reception, as Kuczynski did with her disastrous lip-plumping procedure. There's no real risk that hand transplants will become the latest fad in reconstructive surgery—not least because a dead donor is required. But Clint Hallam's strange search for identity raises interesting parallels with the motivations of cosmetic surgery patients—and instructive contrasts with face transplants, too.

In 1998, when I was Senior Lecturer in Medical Ethics and Law at Imperial College London, I was asked to serve on the clinical ethics committee of St Mary's, Paddington, one of the hospitals associated with Imperial College School of Medicine. Alongside our more routine duties, we were requested to consider the application by an international clinical team—through one of its members, Professor Nadey Hakim, a respected consultant transplant surgeon at St Mary's—to perform the world's first human hand transplant.

Coincidentally, that group also included Professor Dubernard, who was later to lead the team performing the first face transplant.

The case, widely reported and in the public domain, concerned a forty-eight-year-old New Zealander, Clint Hallam, whose arm had been cut off in an accident with a circular saw fourteen years earlier. (Hallam voluntarily revealed his identity in press interviews afterwards, so there's no breach of confidentiality in naming him.) The arm had been reimplanted following the accident but had failed to 'take'. Refusing any of the increasingly sophisticated prosthetic hands available, Hallam then approached the Australian head of the transplant team and asked to volunteer as the first recipient of a human hand 'allograft' (transplant from another person). In turn, the head of the team asked its local member at St Mary's, Professor Hakim, to obtain permission from our recently established clinical ethics committee to perform the procedure there.

In one view, hand transplants crossed technological frontiers at the time, but not ethical ones; the only issues to be resolved concerned professional competence, under the assumption of patient autonomy.[54] Given the presumption of competence to consent in adults, and the lack of any grounds presented to the committee at the time for supposing that the patient lacked capacity, his wishes should have been respected, on this interpretation. It was not up to the committee to judge the risks, but to the patient; anything else would be paternalistic. Or so some members of the committee argued—just as many people think the decision to have cosmetic surgery is entirely a matter of personal choice.

But what if the choice is irrational? Is patient autonomy a catch-all that includes any wish whatsoever? Respecting patient autonomy isn't necessarily the same as giving a patient whatever he wants. It may be wrong to take advantage of another person's willingness to harm himself: motives are complex creatures. Following extensive media coverage of a total artificial heart transplant in 1982, some volunteers were even willing to 'donate' their hearts in the interests of advancing science, although it would kill them.[55] You can't ask a surgeon to do that.

Doctors are people, too, with their own rights and moral beliefs. Is it still part of the doctor's duty to perform a procedure that may actually shorten life, even at the patient's request? Medicine is generally thought to be about improving health and life expectancy, but given the risks inherent in life-long immunosuppressive medication, a hand transplant, like a facial graft, may actually shorten life. Our committee had to navigate uncharted waters in confronting these ethical issues, but we were mindful of the goals and limitations of medicine: 'The art's most delicate aspect is not to shorten life further, and not to diminish it.'[56]

Other innovative transplant procedures, like multi-organ transplants, may have such unacceptably high mortality rates that they are more properly characterised as research than as therapy. (That's also how the surgeon Laurent Lantieri conceives of face transplants: as operations that will further the cause of research.) The difference is that multi-organ transplants are usually intended to save life—indeed, as the last chance to save life. A hand transplant would *not* have been life-saving, we felt, and an artificial hand could actually have performed as well or better than the transplanted hand was expected to do. Given that the benefits of limb transplant didn't outweigh the risks, the autonomy and rationality of the patient weren't necessarily self-evident to us.

In those circumstances, what made Clint Hallam want a hand transplant so badly? That was the question that troubled me, along with the majority of other committee members. We decided that before saying either yes or no, we would like to have a psychiatric assessment of his reasoning and capacity. That has since become standard procedure,[57] but we were proceeding without any precedents. Nor had we had any opportunity to interview Hallam. While we were at pains to emphasise that we were *not* refusing permission for the procedure altogether, we wanted more information about how he saw the risks and benefits.

The clinical transplant team decided to bypass our request and to have the transplant performed elsewhere—in fact, at Professor Dubernard's hospital, in Lyon. In April 1999, the *Lancet* published an Early Report on the initially successful results of this first human hand allograft, performed in September 1998.[58] Meanwhile, the British tabloid press made mincemeat of us for blocking the United Kingdom's chance of medical stardom.

Some months afterwards, Hallam himself turned up at a meeting of the clinical ethics committee, in the company of the transplant surgeon Nadey Hakim, to demonstrate the functioning and appearance of the transplanted hand. The proceedings of that committee are confidential, but suffice it to say that I, for one, was not persuaded that we had been wrong—although at neither committee meeting were we told crucial facts about Hallam's psychiatric history.

Two years later, the *International Herald Tribune* carried an article describing how Hallam had become 'detached' from his hand.[59] This was perhaps an unfortunate turn of phrase. It transpired that Hallam was firmly convinced that the cadaver hand which he had received was really his own. Even before the transplant, he believed that the reason why the hand he lost in the circular saw accident had failed to graft after it was re-attached was that it wasn't actually his to begin with. His own perfect hand—like the 'real me' in the extreme makeover—was out there, just waiting for the surgeon's skills, like the prince hacking through the thorns to rescue Sleeping Beauty.

That conviction had fatal medical consequences. If the hand that Hallam had received in the transplant was 'really' his own hand all along, as he believed, then he wouldn't need to take immunosuppressive medications to prevent rejection, would he? The psychiatric assessment which our committee had requested might have revealed these delusions, but of course it never took place. With the hand showing signs of rejection two years later because of his failure to take his immunosuppressants consistently, Hallam realised, too late, that it wasn't his hand after all. On 3 February 2001, the hand was amputated.

It would take someone with a very firmly rooted identity to cope with waking up every morning to find someone else's hand, a dead man's hand, peeping over the duvet. Even the recipient of an internal organ transplant may face dilemmas about what he owes to the dead donor.[60] How much greater will the constant reminder of otherness be for the recipient of someone else's constantly visible hand? Similar qualms have since arisen in the context of face transplants, where the identity question is even more obvious.

You could see hand and face transplants—unlike cosmetic surgery—as transferring personal qualities from one human being to another. Although Hallam was convinced that he would find his 'real' hand out there, it's more common for recipients of organ transplants to feel that their identities have been somehow invaded by the donor's personality. This goes beyond the question of the hand's visibility, though that too is an issue—a constant threat to the recipient's sense of his *psychological* wholeness, arguably outweighing the *physical* wholeness for which the transplant was sought in the first place. An artificial hand or limb might arguably have the same effect, but on the other hand, there may be a crucial psychological difference. The recipient isn't expected to believe that the artificial limb is either his or another person's, because it's obviously not human. There are no personal qualities to be transposed from one person to another.

The hand is in fact a symbol of connectedness with others: shaking hands, for example, symbolises good faith. There is a wider function of the hand in relation to identity, as an instrument of physical intimacy, of contact with others, of consummate skill in artists and musicians, of agency itself—witness the use of 'hand' to represent agency in phrases such as 'the hand of Fate', 'by his own hand' and 'the hand of God'. The hand plays an unrivalled part in both shaping and standing for the story of the recipient and the donor and in representing agency, and our language reflects this role.

As the physician, poet and philosopher Raymond Tallis has written:

> [T]he hand remains unarguably a bodily structure while at the same time, it has played a crucial role in loosening the bonds that constrain us through our embodied state. The hand—by which we have manipulated, rather than talked, ourselves free of organic constraint—may point the way into the future of mankind.[61]

THE 'REAL ME': WHAT MONEY CAN'T BUY

Personal and human identity is embodied, not abstract. The hand symbolises this embodiment quite strongly, even if not as obviously as does the face. But there's an important difference: the patients on whom face transplants have been performed did have faces, even if damaged or deformed ones. The recipients of hand transplants had lost their hands altogether. Any face, even a less-than-'normal' one, is the primary badge of identity. As James Partridge of Changing Faces says: 'Even for those like myself with severe disfigurement, the face carries a lot of identity, the sense of self.'[62]

Personal identity isn't up for sale: as the Canadian bioethicist Francoise Baylis has written, 'recognition by others, not just recognition of self, is key to identity'.[63] As always, relationship is all-important. The view of face or hand transplants which says that patients should be able to undergo either one entirely as a matter of choice fails to recognise the centrality of relationship—the claims and needs of the donor's or recipient's families, for example. But even more profoundly, it ignores the way in which personal identity is actually created through relationship.

So in a strange, final irony, looking for the 'real me' in a face transplant or cosmetic surgery actually means denying your real identity—just as surely, if not as floridly, as Clint Hallam denied that his real hand was the hand which had originally been re-attached. To insist on attaining some ever-elusive ideal of 'normality' won't confer normality, particularly not when the standards of 'normality' are continually shifted by the cosmetic surgery industry to suit their own profits and purposes. You can't buy the 'real me', although, like Olivia Goldsmith, you can die trying.

8

My body, my capital?

'Shopping for the real me' through cosmetic surgery—like 'body shopping' in general—symbolises both the dream and the nightmare of modern commercial biomedicine. To be more precise, the dream and the nightmare are inseparable, as Alex Kuczynski found to her cost when she herself became a 'beauty junkie'. But even an irrational obsession with cosmetic surgery has a deeper significance: the way in which our bodies are said to have become a kind of capital—our best and truest investment.

As the anonymous woman quoted in Chapter Seven put it, 'All we have in life is ourselves, and what we can put out there every day for the world to see ... Me is all I got.' This book has covered the commercialisation of parts of the body, down to the minute level of genes. In cosmetic surgery, by contrast, it's the *whole* body that becomes an object: the one secure possession, everyone's literal birthright.

So is this woman being realistically hard-headed, or actually exploiting and demeaning herself? Marx, whose influential ideas on exploitation we encountered at the very start of this book, must be turning violently in his Highgate tomb. He thought that it was the propertyless workers, the proletariat, who were distinguished by only having the labour of their bodies to sell, lacking any other capital. But now, it seems, everyone increasingly sees their bodies as their capital.

In his 2005 surprise bestseller, *L'avènement du corps* (*The Coming of the Body*), the French commentator Hervé Juvin actually extols this sort of attitude.[1] Plastic surgery, the implantation of biochips, piercings—all emblazon the new attitude that my body is my unique property. At the same time, because everyone has a body, property has suddenly become democratised, as property in the body.

In an ironic recasting of the sixties feminist slogan, 'Our bodies, our selves', perhaps we ought to say 'My body, my capital'? Juvin thinks that the answer to this somewhat flip question should be a reverberating '*oui*'. We can all make or remake our bodies, enhance our potential wealth in them and improve our chances of a long and healthy life, he argues. We can all be body billionaires.

The long-standing jokes about sexual body parts, 'her assets' or 'his crown jewels', take on a literal and serious meaning, from this perspective. If the body has become everyone's main capital and universal inheritance, then it's worth investing in it. Once we in modern Western society have attained a sufficient level of wealth, enhancement and prolongation of life are what really count. That explanation would fit cosmetic surgery in the United States rather nicely: conspicuous consumption can be best displayed not in your Jimmy Choos, which you only wear some of the time, but rather in your ever-present Hidalgos.

Before dismissing this analysis as jejune and even perverse, it's worth noting the way in which it accords with a more profound way of thinking. We appear to live in a time which has witnessed the absolute failure of the grand Enlightenment dreams of linear progress, universal peace and equality between rich and poor. Together with widespread hostility to organised religion (manifested in popular books such as Richard Dawkins's *The God Delusion*), disappointment in social ideals means that we turn inwards. In the absence of a belief in eternal life, everything becomes invested instead in this life, in this body. Long life is our ferocious desire, eternal youth our supposed right and the myth of the body without origin or limits our crusading new religion.

That could also mean that we look to the state to maintain our bodily investment. The state's job is then to guarantee to each of us the healthy growth of our investment in a healthy body. That may be why governments are so widely seen to have a positive duty to promote stem cell research and other forms of medical progress. The biotechnology industries flourish, with state sanction and support, because they add extra value to the body, which is already the object of supreme worth to us. (Woe betide any bioethicist who stands in the way of these combined forces of state, industry and deep popular longing.)

As noted earlier, the body has become accustomed to prostheses and aids that liberate it from memory, time and space. The infinite renewal of the body isn't confined to superficial repairs through cosmetic surgery. It includes the way in which external substitutes like pacemakers can be surgically incorporated into the body, breaking down the barriers between the body and the outside world. At the same time, tissue removed from the body enters into external commerce and trade, as a commodity like any other.

What was once termed the 'End of History' or the 'End of Ideology', by thinkers as disparate as Hegel, Marx and the neo-conservative writer Daniel Bell,[2] is, in this interpretation, the 'End of Nature'. Rather than a new classless society rendering class warfare irrelevant, or a democratic ideology emerging victorious over all opposition, it is biotechnology that has come out triumphant. We need no longer accept the merely natural body. We can do better than that.

For the fortunate few who can afford these procedures to stave off decay and death—although obviously not for good—perhaps the 'End of Nature' is an attractive notion. But isn't Juvin's picture an otherwise too grandiose and exaggerated view of bioindustry and biopolitics? Yes: despite his compelling rhetoric, Juvin has fallen for the same sort of frenetic 'hype' which predisposed commentators to believe that Hwang Woo-Suk really had discovered the Fountain of Youth, in the form of therapeutic cloning.

Such a highly optimistic analysis obviously fails on the clinical front: cord blood spare part kits don't work as well as allotransplants, face grafts require lifelong immunosuppressants and cosmetic surgery can kill rather than prolong youth. Medically, this grand thesis is glibly improbable, but it's also criminally naïve economically. Juvin may live in a cosy world where the body is the only capital that counts because everyone has sufficient other wealth, but the vast majority of humanity doesn't. Here are just two out of many possible examples.

ORGANS FOR SALE, ONE CAREFUL (AND UNWILLING) OWNER

Since the tsunami of 2004 further impoverished the already desolate fishing villages on the Bay of Bengal, fifty-one women from one village alone have sold a kidney, to rescue their families from an equally threatening tidal wave of debt. Most 'donors' are lower-caste women in their twenties; most recipients, high-caste Indians or wealthy Westerners. Stranded in refugee villages miles from their fishing boats, the women's husbands can't feed their families. So the burden of putting the food on the table falls on their wives, who have few other means by which to do so other than selling their kidneys. The going price is about 35,000 rupees (£460), with a substantial percentage going to the local broker.

Dr Ravindranath Seppan, a campaigner against the cash-for-kidneys trade, claims that this village is by no means unique. He cites an article in the *Journal of the American Medical Association* which surveyed 305 Indians who had sold a kidney to pay off their debts. Three-quarters of them remained in debt, but now with worsened health. A weak climate of government regulation, despite laws on the Indian statute books prohibiting cash for kidneys, allows the trade to continue.[3] World Health Organisation guidelines issued in 1991 also ban payment for organ donation, but although 192 countries have endorsed them, they aren't binding.

Often the debts that the women are trying to repay are health-related, which can be expected to become commoner as governments limit socialised medicine or privatise hospitals, as part of structural adjustment plans imposed by the International Monetary Fund or World Bank. But of course this will become a vicious circle: when these weakened women need health care themselves, who will sell a kidney to pay for them?

Capital is money used to breed more money, but selling kidneys out of desperation only breeds further poverty. It seems worse than insensitive to typify what these Indian women do as making capital from their bodies, even though there is some vestigial element of choice in their actions. But that minimal autonomy doesn't apply to the victims of the Chinese organ trade, who are largely members of the despised Falun Gong Buddhist sect or prisoners condemned to death for any of the 160 capital offences in China. Nor is this trade confined to kidneys, of which we're all born with a 'spare'. The 'capital' in prisoners' bodies extends to their hearts, lungs and livers. The only snag is that you can't live without those organs.

Although cynics might shrug off tales of cash-for-hearts as mere urban myths, the use of condemned prisoners as cadaver 'donors' for the international organ trade was openly acknowledged by the Chinese deputy health minister, Huang Jiefu, in November 2006. Amnesty International had been reporting large-scale 'harvesting' of vital organs from prisoners since 1993. In 1998, the European Parliament passed a resolution condemning the sale of organs from executed prisoners. For many years, international medical bodies such as the British Medical Association had also condemned the practice.[4] But no Chinese medical disciplinary body exists to impose standards; what the state permits goes, and the state not only permits the organ trade—it actively encourages it. Coincidentally or not, after the state began its persecution of Falun Gong in 1999, the number of liver transplants surged from 118 to over three thousand in the space of four years.

A report by a Canadian human rights lawyer, David Matas, and a former Canadian Member of Parliament, David Kilgour, gave further substance to allegations about the Chinese organ trade, even specifying the going rates for particular organs in the global transplant market:[5]

Kidney US$62,000
Liver $98,000–130,000
Liver-kidney $160,000–180,000
Kidney-pancreas $150,000
Lung $150,000–170,000
Heart $130,000–150,000
Cornea $30,000

The Canadian report found that the global trade in prisoners' organs has become a vital source of hard currency income for the Chinese health and military systems, now that the prop of state funding has been kicked away. As Matas and Kilgour wrote: 'The organ harvesting market in China, in order to thrive, requires both a supply and a demand. The supply comes from China, from prisoners. But the demand in large part, in big bucks, comes from abroad.'[6]

Chinese hospital websites compete to offer the shortest waiting times—at one to two weeks, suspiciously lower than anywhere else in the world—for their busy international clients. These 'foreign friends' are offered the promise that 'providers can be found immediately!' (original exclamation marks). How is this miracle of modern medicine possible? The website goes on to assure 'foreign friends' that:

> So many transplantation operations are owing to the support of the Chinese government. The supreme demotic court, supreme demotic law-officer, police, judiciary, department of health and civil administration have enacted a law together to ensure that organ donations are supported by the government. This is unique in the world.[7]

A website affiliated to the Number Two Military Medical University offers a rock-bottom price of 200,000 yuan (£7,200) for a liver transplant, hospitalisation and operation fee—with the numbers of transplants having skyrocketed in recent years. The funding which military hospitals receive from tourists who jet in to buy organs not only keeps the hospitals themselves afloat; it subsidises the rest of the military as well. (Buy yourself a kidney, keep the Chinese occupation of Tibet going ...)

Although a Chinese law against organ sales came into belated force in July 2006, the Canadians remarked caustically that:

> ... from what we can tell, the law is not now being enforced. Belgian Senator Patrick Vankrunkelsven, in late November 2006, called two different hospitals in Beijing pretending to be a customer for a kidney transplant. Both hospitals offered him a kidney on the spot for 50,000 euros.[8]

Nor are transplant tourists subject to prosecution in their own countries if it turns out that the organ was obtained without the donor's consent, even by the murder of the 'donor'. While a sex tourist who uses child prostitutes abroad can sometimes be prosecuted at home for child abuse, transplant law is strictly territorial; it hasn't woken up yet to the worst abuses of global body shopping.

These two examples demonstrate how hopelessly inadequate it is simply to view the body as an individual's private capital. For the Chinese state, the organs of executed Falun Gong members and prisoners are part of the nation's assets. And for these unfortunate people, the 'biovalue' in their bodies doesn't liberate them; instead, it puts them at mortal risk.

THE TRAGEDY OF THE GENETIC COMMONS

It's clearly misguided to think of 'body shopping' as an entirely beneficent phenomenon, in which we all blissfully increase our capital in our bodies. But how else might we understand it? Two better models will be explored in the rest of this final chapter.

The American law professor James Boyle believes that we can best grasp the way in which the body has become an object of trade by likening it to the historical process of enclosure. In Britain, during the great agricultural transformation in the eighteenth century that preceded the industrial revolution, land, which had previously been a public resource, was 'enclosed' by private landowners. Freed of feudal-style legal restrictions on transfer of ownership and of traditional rights held by commoners who used communal land to pasture their animals, landholdings could now be sold to raise capital, which helped to fund the industrial revolution.

As in the old agricultural enclosure movement, so in today's biotechnology: 'things that were thought to be uncommodifiable, essentially common or outside the market altogether, are being turned into private possessions under a new kind of property regime'.[9] In biomedicine, a series of legal cases, like *Moore* and *Greenberg*, have generated a similarly powerful momentum towards the transfer of rights over the body and its component parts from the individual 'owner' to corporations and research institutions. The body has entered the market, just as land did when it was turned from a public good to a private one. So in a sense, it *is* correct to argue that the body has become capital, just as land did—but not to blithely assume that everyone benefits, any more than the dispossessed commoners grew wealthy during the agricultural enclosures.

The metaphor of 'body shopping' as a form of enclosure works well for many of the developments this book has examined. The promise that every baby can be equipped with a personal spare parts kit, for example, is promoted heavily in advertising for private cord blood banks. That's a prime example of how something that used to be outside the market altogether, cord blood, has become the subject of burgeoning market competition, and also a new kind of privatisation.

The enclosure metaphor also helps us to analyse resistance to 'body shopping': instances in which it's withstood as much as those in which it's allowed to go ahead. Although he doesn't extend it that far, Boyle's argument would fit the reasoning that the French national ethics committee used in rejecting private cord blood banks:

> Preserving placental blood for the child itself strikes a solitary and restrictive note in contrast with the implicit solidarity of donation. It amounts to putting away in a bank as a precaution, as a biological preventive investment, as biological insurance ... There is major divergence between the concept of preservation for the child decided by parents and that of solidarity with the rest of society.[10]

By continuing to insist that cord blood and other forms of human tissue are a *public* resource, the official French view defends an older, pre-enclosure model of society: less individualistic, more organic, less infatuated with the notion of *private* choice.

Boyle intends his enclosure metaphor to apply widely, from computer software to human spleens—specifically, John Moore's.[11] His target is 'the relentless power of market logic to migrate to new areas, disrupting traditional social relationships, views of the self, and even the relationship of human beings to their environment'.[12] But the enclosure metaphor fits genetic patenting particularly snugly.

Just as the agricultural enclosures turned common wealth into private possessions, so the 'great genome grab' attempts to profit from and privatise the public resource of the human genome—both in the Third World (as in the Tongan case) and in the developed countries (as in the 'French DNA' example). Far from providing all of us with capital in our bodies, this process transfers wealth from the public domain to biotechnology venture capital and start-up firms—where it succeeds, as it didn't in Tonga and France.

When the agricultural poor had pre-existing rights in the public resource, like grazing for their animals, they lost out. As the popular rhyme of the period went:

> The law will hang the man or woman
> Who steals the goose from off the common,
> But lets the greater villain loose
> Who steals the common from the goose.

And the poor lost out despite the codified rights of commons which the law had granted them since feudal times: rights such as grazing, coppicing, water and footpath access. What hope do we have of protecting the genetic commons from similar depredations, when the common law has never recognised our rights to property in our bodies? As we've seen over and over, the traditional legal position held that tissue, once excised from the body, has either been abandoned as waste by its original 'owner' or that it was 'no one's thing', which never belonged to anyone in the first place.

Appropriation of the genetic commons has been justified by another argument, Garrett Hardin's influential notion of 'the tragedy of the commons'.[13] In Hardin's view, communal property is prone to abuse, because everyone who has common rights in an object has an incentive to overuse it. As a whole, the community loses out when individuals take more than their share, but it's in each single person's interest to use the common resource as intensively as possible, to become a 'free-rider'. The defenders of land enclosures argued that private ownership would avoid the tragedy of the commons by eliminating incentives for overuse, transferring inefficiently managed common land into single ownership.

Even though it's hard to see how ordinary individuals could possibly 'overuse' the genome, a similar argument is frequently made about more efficient uses through private rights like patents. Advocates of corporations' control of the new biotechnologies can frequently be heard claiming that property rights for them will mean long-term benefits for all. What's good for Myriad Genetics is good for the *world*, in this view. Even General Motors more modestly claimed that what was good for it, was good for the *nation*.

Actually, the resource of land was often grossly mismanaged by the new private owners—as when Scottish crofters were evicted in the nineteenth century to make way for sheep, just at the time when the bottom was about to fall out of the British wool market. Emigration nearly destroyed the Gaelic language and left large areas radically underpopulated to this day. Uncultivated land quickly sank back into peat bog. The failures of sheep-farming gave way to the conversion of many large estates to deer parks for shooting purposes. When these in turn fell out of favour—large Victorian house parties no longer being the thing after World War I decimated the population of footmen and valets—the deer bred too rapidly. As a result, it's now being debated whether the grey wolf should be reintroduced to the Highlands.

In fact the Scottish example actually illustrates 'the tragedy of the *anti*-commons':[14] a private right to prevent others from using property of mutual interest, resulting in *underuse* of the monopolised resource. On the island of Raasay, near Skye, enclosure of the productive southern two-thirds of the island drove the remaining population into the barren northwest section and the nearby rocky island of Rona, bereft of pasture and arable land. A six-foot-high stone wall was built across the narrow northern neck of Raasay to keep the sheep and deer in, but the crofters out. The owner of the newly enclosed estate, a merchant called George Rainy, even enforced a rule that no one should marry on the island, to reduce the remaining population by natural attrition. In the 1841 census, nearly one thousand people had lived on Raasay, almost all in the southern and eastern townships. Twenty years later, only deserted villages like Hallaig, evoked in the Raasay poet Sorley MacLean's poem of that title, remained of 'the green and cultivated land ... on the tops of the high eastern cliffs, which [were] everywhere covered with farms.'[15]

Likewise, when a gene is patented defensively, the patentholders can block research and treatment that would benefit everyone. That certainly happened in the *Greenberg* and *Catalona* cases, as well as in the example of the BRCA1 and BRCA2 genes. Another instance of the tragedy of the anti-commons is the drug Herceptin, which has innovative therapeutic uses against cancer cell production, in women with certain genetic predispositions to breast cancer.

Herceptin acts on the HER-2 gene (human epidermal growth factor receptor-2) and increases survival rates in women who have the version of the gene that makes them more prone to some forms of breast cancer. (About 25 per cent of all breast cancers are HER-2 positive.) The patentholder of Herceptin, Genentech, also holds multiple patents related to the HER-2 gene *itself*. Any researcher or drug company wishing to develop an alternative, cheaper drug must obtain permission from Genentech or risk being sued for patent infringement.[16] This monopoly has driven the price of the drug up to such high levels that the UK Institute for National Clinical Excellence had to restrict its use on the National Health Service, in a controversial and widely unpopular decision.

The Herceptin example illustrates the concrete threat to both research and treatment from privatisation of our 'commons' in the human genome and bodily tissue. Usually, of course, defenders of commercialisation present it instead as the helpmate of research and therapy. Boyle is particularly scathing about the disingenuousness—

accepted too readily by the courts in cases like *Moore*—of corporate claims that granting individuals a property right in their own tissue would hinder research and treatment: 'To back up this argument, the court [in *Moore*] paints a vivid picture of a vigorous, thriving public realm. Communally organized, altruistically motivated and unhampered by nasty property claims, the world of research is moving dynamically toward new discoveries.'[17]

In order to justify its claims for private property rights, the biotechnology sector has been allowed by the courts to claim that it alone represents the public interest—to which it is arguably the greatest *threat*. The *Moore* court accepted that granting Moore and others like him a property right in their own tissues would only encourage selfish claims. Compulsory altruism was to apply in the strictest possible fashion to donors, but strictly only to donors. And one-way altruism, as I've pointed out elsewhere, is better termed exploitation.[18]

Those who favour giving individual patients private rights to buy and sell organs and tissue are on the firmest ground when they highlight the hypocrisy of a 'gift' relationship in which only the donor is expected to demonstrate selflessness. When the UK Human Fertilisation and Embryology Authority overlooks the real risks of egg donation in encouraging women to show their altruism, or when women are asked to endure additional risk in delivery for the sake of providing cord blood to profit-making banks like Branson's, the gift relationship is similarly abused.

In a passage whose rhetorical final question is clearly intended to be answered in the same vein as queries about the doctrinal allegiance of the pontiff or the lavatorial habits of bruins, Boyle likewise declares:

> On the one hand, property rights given to those whose bodies can be mined for valuable genetic information will hamstring research because property is inimical to the free exchange of information. On the other hand, property rights *must* be given to those who do the mining, because property is an essential incentive to research. Do these assertions contradict each other?[19]

Actually, perhaps this pair of questions is a little tougher than the ones about the pope or the bears. Those who want to bestow greater property rights on researchers and the corporations or universities that employ them would probably argue that the contribution made by researchers adds more value to the finished product. Those who want to extend property interests to mere 'raw material' providers, like women who donate ova for stem cell research, might counter, as I have, that their labour is undervalued, although it also contributes vital worth.

Even though I argued that 'to pay or not to pay' was not the real question about egg provision, I did also make the Lockean claim that women put extensive labour into donating eggs and deserve recognition for it. They put hours of work into 'donating' and take risks of the sort that are supposed to justify rewards, in the case of entrepreneurs or venture capitalists.

Boyle thinks that the 'raw material' providers like Moore—he doesn't discuss egg donors—typically do find their contribution either undervalued, or not valued at all. This, he says, is because human tissue represents a public commons, and because our dominant political culture sets the private above the communal. Not only does it value private property more highly; it also asserts that private wealth actually produces public wealth. The public sector is thought to be dependent, or even parasitic, on the private. (That's the implicit assumption behind the attempt to impose market models of finance and management on the public sector in the United Kingdom during the Thatcher, Major and Blair years: 'private good, public bad'.)

As Boyle points out, Marx thought the reverse. 'Writing about the industrial revolution and the transformation of capitalism, Marx turned the rhetoric of private property and entrepreneurialism on its head, arguing that wealth was socially produced but privately appropriated.'[20] In the agricultural enclosure movement, as Boyle presents it, we see the wellspring of the industrial revolution: the massive appropriation of the public commons in land. That earlier 'great grab' provided both the capital to invest in new machines and the workforce to operate them: it forced the rural British poor into urban factories, since they no longer had any entitlements to communal resources in the countryside. Now it's the *global* poor who are forced to sell their kidneys because they have no other resources.

The commodification of human tissue in 'body shopping' also expropriates common resources—which is exactly how the Tongans conceive of their genetic heritage—and turns them into private wealth. Boyle highlights similar Third World examples involving plants, rather than human tissue. For instance, *vinca* alkaloids from the rosy periwinkle of Madagascar, long used by Malagasy peoples to treat diabetes, have now yielded a drug patented by Eli Lilly—worth $100 million a year for the treatment of a different condition, Hodgkin's disease. If Madagascar had received any share in this income, it would have been one of the country's largest sources of wealth. As it is, local people have been reduced instead to chopping down forests for slash-and-burn agriculture.[21]

So Boyle's enclosure metaphor encompasses a wide range of examples of 'body shopping', helping us see common links between the disparate forms it's taken in this book. There's one thing his model doesn't do, though, and that's to explain the way in which women's tissue has been a particular focus of commodification. Nor does it really scrutinise the popular fear and loathing that seem to be attached to so many forms of 'body shopping'. For that, we need to look, in this final section, at a different argument: my own view that we all have female bodies now.

WHY WE ALL HAVE FEMALE BODIES NOW

Whereas Juvin celebrated the way in which the body has become an object of trade, it seems to me that the opposite reaction is much more widespread. Most people are quite shocked when they learn that Moore wasn't held to have a property in his body, or that one-fifth of the human genome has been patented, mostly by private firms. But to play devil's advocate, why should they be so surprised? After all, female bodies have been subject to various forms of property-holding over many centuries and in many societies.

Women's bodies are used to sell everything from cars to pop music, of course; there's nothing remarkable or original in making that point. But female tissue has been objectified and commodified in much more profound ways, in legal systems from Athens onwards. Men were also made into objects of ownership and trade, of course, under slavery. But women were much more likely than men to be treated as commodities in non-slave-owning systems. Even though women weren't literally the property of their husbands under the

Anglo-American legal system of 'coverture' governing marriage, for example, their husbands had the right to own and manage their income, their capital and their labour.[22] Men held many property rights over women's activities, even if they didn't actually own their wives as slaves.

And once a woman had given her initial consent to the so-called marriage 'contract', she had no right to retract her consent to sexual relations at any time. Rape in marriage wasn't recognised as a crime, because the husband was presumed to have a sort of property right over his wife's body, at least for the purposes of sexual relations. Usually contracts protect both parties' property rights, but in the case of the marriage contract, the enforcement mechanisms worked almost entirely in one party's favour: the husband's.

It doesn't take a great leap of the imagination to see the parallels between that situation and the way in which the law offered little redress to Moore or Greenberg. Similarly, when pathologists like van Veltzen or employers like British Nuclear Fuels simply take the organs of dead children or employees at autopsy without consent, *all* bodies are being treated as open-access, like women's bodies under the marriage 'contract'. Just as the language of women's natural capacity for love and selfless devotion has been used to hide the power imbalance in marriage, so the language of 'gift' is employed to camouflage one-way altruism in relationships between donors and biobanks.

In that cradle of democracy, ancient Athens, freeborn women weren't slaves, but they had no title to any property whatsoever, not even their own clothes. An Athenian woman wasn't a party to her own marriage contract: that was a transaction between her father as her present *kyrios* (lord) and her husband-to-be as her future one. Although modern commentators have often been volubly shocked by the lack of freedom of choice in Plato's proposals for eugenically dictated marriages in his *Republic*, they ignore the fact that no Athenian woman had a free choice of whom to marry, or indeed whether to marry at all.

What we see in such outrage is a similar phenomenon to the feminisation of the body in modern biotechnology: the assault on freedom is only noticed when it begins to apply to men. It took a very long time for people to notice that women's eggs were required in large quantities for the stem cell technologies: the phenomenon I've termed 'the lady vanishes'.[23] The stem cell debates seemed to be premised on the assumption that only the status of the embryo mattered.[24] Many people are still quite unaware that women's eggs are a crucial part of 'therapeutic cloning'; the vituperative 'cloning wars' rarely mention that issue. By contrast, genetic patenting, which affects both sexes, has generated a huge scholarly literature and a very vigorous popular debate. Just a coincidence?

I think not. The taking of solely female tissue doesn't provoke widespread concern; rather, we hear constant calls for women to be more altruistic, trust the scientists and stop making a fuss about the risks—very much the tone of the HFEA decision in February 2007 to allow egg donation for research. Yet it's often female tissue that possesses the greatest 'biovalue'. And the processes by which it's 'harvested', at least in the case of eggs for stem cell research, are much more invasive than the taking of a cheek swab for genetic analysis. In the case of Catalona's tissue donors, no further procedures were required at all: the men had already undergone their operations, from which the samples were left over. (Perhaps it's also no coincidence that the plaintiffs in the Catalona case, which attracted widespread media outrage, were by definition all men, donors of prostate tissue samples.)

My original argument—provocative, if you like, but also productive, I hope—is that what we are witnessing is *fear of feminisation of property in the body.*[25] The 'new enclosures' of the genetic commons or of body tissue threaten to extend the objectification and commodification of the body to both sexes. Everyone has a 'female' body now, or, more properly, a feminised body. While men obviously don't have biologically female bodies, both male and female bodies are increasingly at risk of being treated like objects and commodities.

From the time of the Greeks until the eighteenth century, all bodies were regarded as essentially male. Female bodies were seen as imperfect versions of male ones, and women as anatomically failed men. Ovaries were labelled 'female testicles' in diagrams, and the vagina was depicted as an inverted penis.[26] What women contributed to gestation was supposedly no more than what the earth does in harbouring the seed. Their bodies were merely seen as inert matter, while the male 'generative principle' provided the life-giving spark. (In similar fashion, in many accounts of 'therapeutic cloning' an enucleated egg is merely a receptacle for the energising power of the somatic cell transferred into it.)

This one-sex-fits-all model was supplanted after the Scientific Revolution of the seventeenth century by an equally implacable division into two sexes, whose twain would never meet. Now modern biomedicine is reverting to a single-sex model—but with the crucial difference that all bodies are regarded as essentially female, open and accessible. Human tissue, both men's and women's, becomes the mere matter on which biotechnology performs its masculine life-giving magic.

That model has its limitations. Women's tissue is still generally more valuable in 'body shopping'. Objectification and commodification continue to be perceived as more 'normal' for women's bodies. Women's bodies seem to be more widely assumed to be 'open access'. The only difference from past practice is often that the object or commodity takes new and sometimes disturbing forms, like the documented use of aborted tissue from pregnant Ukrainian women to create 'beauty treatments' in Moscow salons,[27] or the trafficking of women for both sex and egg sale in Cyprus clinics.

But when men's bodies likewise become mere objects or commodities, people take notice. The 'new enclosures' threaten both sexes, although they don't affect them equally: female tissue is generally more valuable, as in the case of eggs used in 'therapeutic cloning'. There are exceptions, like John Moore. But then there was also Henrietta Lacks.

I certainly don't want to argue that because women's bodies have traditionally been treated in ways that befit objects rather than people, so should men's bodies be. Equal misery for all is not a very attractive rallying cry. Rather, a feminist approach like mine gives us the historical awareness and the present-day sensitivity we need to protect both sexes. For example, the Korean feminists who were suspicious of Hwang's initial claims, and who asked awkward questions about where that huge number of eggs could have come from, helped to bring the abuses of his research to light. As Paik Young-Gyung said, 'If you argue that it was feminists who helped open up the scandal, I think this is very much true to the case.' Their dissidence was uncomfortable and courageous, at a time when other commentators were heaping praise on Hwang and salivating at his 'enviable supply of eggs'.[28] Korean WomenLink and Solidarity for Biotechnology Watch raised important issues about how we can protect those who are merely 'raw material providers', rather than collaborating research subjects.

Those questions will arise in greater and greater number, in a wider and wider range of countries, it's fair to predict. We urgently need to find better ways of dealing with them: new practical models of governance like the biobank as a charitable trust, and new conceptual understandings like the enclosure and feminisation models. Feminists or not, we all need to jettison the old metaphors about bodies as merely *things*, to be appropriated at will.

Unfortunately, treatment of the new biotechnologies in the popular media and in academic literature alike has actually reinforced such oversimplifications. That lack of imagination isn't just an inevitable reaction to the 'weirdness' of the new biotechnologies. Just because the new biotechnologies are novel doesn't mean that the underlying ethical problems and political phenomena are utterly beyond our previous experience.

We've already seen that the historical enclosure of the agricultural commons provides useful insights into what's going on in genetic patenting. Feminisation of the body is another such comparison or metaphor, invoking another set of historical examples.[29] If we can understand this history, we aren't doomed to repeat it. The two interpretations I've offered here—enclosure and feminisation—can work in tandem. We don't have to choose one or the other. Feminisation makes up for the gender-blindness of enclosure, while enclosure suggests economic arguments about private and public wealth that feminisation misses.

What works much less well is the smug and complacent view that we're all growing rich on the capital in our bodies. Even less convincing, to my mind, is the common assertion that the commercialisation of the body is just a matter of free individual choice. And the least convincing of all, I think, is the claim that anyone who opposes 'body shopping' is a stick-in-the-mud enemy of medical and scientific progress—when commercialisation of the genome and human tissue is so often actually an *obstacle* to research and therapy.

'Body shopping' sometimes seems entirely new. Many of its worst abuses—the theft of Cooke's bones, the murder of Falun Gong believers for their organs, the trafficking of Eastern European women for their eggs—are genuinely shocking. We shouldn't lose that sense of dread and awe; its heart, so to speak, is in the right place. But neither should we conclude that the commercial forces behind 'body shopping', and the onward momentum of modern biotechnology, are so overwhelming as to be unstoppable.

The modern French philosopher Maurice Merleau-Ponty wrote, 'For if the body is a thing among things, it is so in a stronger and deeper sense than they.'[30] When genes are patented, eggs are 'harvested', or cord blood is 'banked', that strangeness is ignored. The body both is, and is not, the person. But it should never be only a consumer good, an obscure object of material desire, a capital investment, a transferable resource: merely a thing. Our consciousness, dignity, *ngeia* and human essence are all embodied, caught up in our frail human bodies. The body is indeed like nothing on earth: not no one's thing, but no thing at all.

Endnotes

Preface

1. Ed Pilkington, 'A state of ill health' (2007) *Guardian*, 13 June, section 2, p. 6.

Chapter one

1. Debora Spar, *The Baby Business: How Money, Science and Politics Drive the Commerce of Conception* (Cambridge, MA: Harvard Business School Press, 2006), p. xii.
2. The Spanish Register of Fertility Clinics, last updated 1 February 2003, registered 203 authorised centres for the practice of reproductive medicine; 38 were public and 165 private. Among the public centres, only seven performed oocyte donation, and none of them paid donors for their eggs. My thanks to Prof. Itziar Alkorta Idiakez for these statistics.
3. Susan Weidman Schneider, 'Jewish women's eggs: a hot commodity in the IVF marketplace' (2001) 26 *Lilith*, 3, 22. Schneider claims to have uncovered a flourishing market in Jewish women's eggs, which command a premium because Jewish descent is accounted to be matrilineal, and Jewish women are statistically more likely to pursue professions and to defer childbearing.
4. Spar, *The Baby Business*, p. xvi. A survey by the American Society of Reproductive Medicine in 2007 showed that the average price among responding clinics was $4,217, but over half of the 394 clinics surveyed failed to respond. One might surmise that those clinics who paid more than the ARSM's recommended maximum of $5,000 would be likely to keep quiet.
5. Allen Jacobs, James Dwyer and Peter Lee, 'Seventy ova' (2001) 31 *Hastings Center Report*, 12–14.

6. Spar, *The Baby Business*, table 1.1, p. 3, based on data provided by the American Society for Reproductive Medicine, the Centers for Disease Control, Business Communications Company and individual providers.
7. Anne Pollock, 'Complicating power in high-tech reproduction: narratives of anonymous paid egg donors' (2003) 24 *Journal of Medical Humanities*, 241–63. How accurate these facts are is unclear: one woman interviewed by Pollock deliberately omitted her favourite hobby, karate, for fear of being thought too unfeminine by would-be buyers.
8. Cited in Rob Stein, '"Embryo bank" stirs ethics fears', *Washington Post*, 6 January 2007, p. A01.
9. Ibid.
10. Quoted in Yahoo Daily News, 14 March 2007, 'Gay male parents get dedicated fertility program'.
11. Spar, *The Baby Business*, table 2.2, p. 55, based on data for 2001 from the Centers for Disease Control and Prevention, Division of Reproductive Health.
12. Spar, *The Baby Business*, p. 50.
13. Antony Barnett and Helena Smith, 'Cruel cost of the human egg trade', *Observer*, 30 April 2006, pp. 6–7, at p. 7.
14. Spar, *The Baby Business*, p. 46.
15. Barnett and Smith, 'Cruel cost of the human egg trade', p. 7.
16. Quoted in Barnett and Smith, 'Cruel cost of the human egg trade', p. 7.
17. Barnett and Smith, 'Cruel cost of the human egg trade', p. 6.
18. Suzi Leather, quoted in Barnett and Smith, 'Cruel cost of the human egg trade', p. 7.
19. Among others, this view is taken by the British bioethicists John Harris ('An ethically defensible market in human organs', with C. Erin [2002] 325 *British Medical Journal*, 114–15), Julian Savulescu ('Is the sale of body parts wrong?' [2003] 16 *Journal of Medical Ethics*, 117–19), Janet Radcliffe Richards ('The case for allowing kidney sales' [1998] 352 *Lancet*, 1950–2) and Stephen Wilkinson (*Bodies for Sale: Ethics and Exploitation in the Human Body Trade*, Routledge, 2003), as well as the American academics Kieran Healy *(Last Best Gifts: Altruism and the Market for Human Organs*, University of Chicago Press, 2006), Michele Goodwin (*Black Markets: The Supply and Demand of Body Parts*, Cambridge University Press, 2006) and Mark Cherry (*Kidney for Sale by Owner: Human Organs, Transplantation and the Market*, Georgetown University Press, 2005). For the opposite view, see the American commentators Cynthia B. Cohen ('Public policy and the sale of human organs', *Kennedy Institute of Ethics Journal*, 12, 2002, pp. 47–67) and Lori Andrews and Dorothy Nelkin (*Body Bazaar: The Market for Human Tissue in the Biotechnology Age*, New York, Crown, 2001), as well as my own

Property in the Body: Feminist Perspectives (Cambridge: Cambridge University Press, 2007), and Catherine Waldby and Robert Mitchell's *Tissue Economies: Blood, Organs and Cell Lines in Late Capitalism*, Duke University Press, 2006).

20. Alireza Baghan, 'Compensated kidney exchange: a review of the Iranian model', *Kennedy Institute of Ethics Journal*, 16, 2006, pp. 269–82.
21. Alfred Cohen, 'Sale or donation of human organs', *The Journal of Halacha and Contemporary Society*, 352, 2006, pp. 37–67.
22. Cecile Fabre, *Whose Body Is It Anyway? Justice and the Integrity of the Person* (Oxford: Oxford University Press, 2006).
23. Ibid., p. 133, original emphasis.
24. M. Goyal et al., 'Economic and health consequences of selling a kidney in India' (2003) 288 *Journal of the American Medical Association*, 1589–93.
25. Ruth-Gaby Vermot Mangold, *Trafficking in Organs in Europe*, report no. 9822 (Council of Europe, 2003).
26. Spar, *The Baby Business*, p. xvi.
27. Ibid., p. 39.
28. Carolyn McLeod and Francoise Baylis, 'Feminists on the inalienability of human embryos' (2006) 24 *Hypatia* 1–24.
29. See, for example, Jeffrey Kahn, 'Can we broker eggs without making omelets?' (2001) 1 *American Journal of Bioethics*, 14–15; Gregory Stock, 'Eggs for sale: how much is too much?' (2001) 1 *American Journal of Bioethics*, 26–7; Bonnie Steinbock, 'Payment for egg donation and surrogacy' (2004) 71 *Mount Sinai Journal of Medicine*, 255–65; Mark V. Sauer, 'Indecent proposal: $5,000 is not "reasonable compensation" for oocyte donors [editorial]' (1999) 71 *Fertility and Sterility*, 7–8; Josephine Johnson, 'Paying egg donors: exploring the arguments' (2006) 36 *Hastings Center Report*, 28–31; and Charis Thompson, 'Why we should, in fact, pay for egg donation' (2007) 2 *Reproductive Medicine*, 203–9.
30. Kimi Yoshino, 'Fertility scandal at UC Irvine is far from over', *Los Angeles Times*, 22 January 2006. My thanks to Diane Beeson for providing me with this reference.
31. Commercial tissue supplier Augie Perna, quoted in Annie Cheney, *Body Brokers: Inside America's Underground Trade in Human Remains* (New York: Broadway Books, 2006), p. 189.
32. However, the van Veltzen case did culminate in the revelation that over 54,000 body parts, stillborn babies and foetuses retained for post mortem examination since 1970 were still held by pathology services in England. (UK Department of Health, *Human Bodies, Human Choices: The Law on Human Organs and Tissue in England and Wales, a Consultation Report*, DOH, 2002)

33. Rachel Williams, 'Four accused of stealing Cooke's bones', *Guardian*, 24 February 2006, p. 1.
34. Mark Honigsbaum, 'Hospitals refuse to warn of bone contamination', *Guardian*, 6 January 2007, p. 4.
35. Gary Younge and Sam Jones, 'Family's dismay after Alastair Cooke's bones stolen by New York gang', *Guardian*, 23 December 2005, p. 3.
36. Susan Cooke Kittredge, 'Dad and the bodysnatchers', *Sunday Times*, 12 March 2006, p. 3.
37. Cheney, *Body Brokers*, pp. 187–8.
38. Ibid., p. 9.
39. Ibid., p. 184.
40. Tom Parfit, 'Beauty salons fuel trade in aborted babies', *Guardian Unlimited*, 17 April 2005, online at www.guardian.co.uk. The report alleged that women were paid extra to have late abortions, since foetuses at an advanced stage of development were thought to have greater restorative powers. In a context where abortion was, until recently, the normal mode of 'contraception', vulnerable women may feel fewer qualms about this procedure; corrupt doctors, it is alleged, are even advising women to have a termination on grounds of foetal abnormality where none exists. An illicit trade between Ukraine and Russia provides the foetuses to Moscow beauty salons, where they are sold for up to £5,000 each.
41. Nick Madigan, 'Inquiry widens after two arrests in cadaver case at UCLA', *New York Times*, 9 March 2004, p. A21.
42. Cheney, *Body Brokers*, p. 156.
43. Ibid., p. 8.
44. Ibid., p. 31.
45. Ibid., p. 63.
46. Interviewed by Cheney in *Body Brokers*, p. 69.
47. Interviewed by Cheney in *Body Brokers*, p. 75.
48. Interviewed by Cheney in *Body Brokers*, p. 78.
49. Lesley A. Sharp, *Bodies, Commodities and Biotechnologies: Death, Mourning and Scientific Desire in the Realm of Human Organ Transfer* (New York: Columbia University Press, 2007), p. 56.
50. Letter from Rogers to Brown, cited in Cheney, *Body Brokers*, p. 78.
51. Cheney, *Body Brokers*, p. 9.
52. Jill McGivering, 'China "selling prisoners' organs"', BBC News online, 19 April 2006; Edward McMillan-Scott, 'Secret atrocities of Chinese regime', *Yorkshire Post*, 13 June 2006; Ethan Gutmann, 'Why Wang Wenyi was shouting: is Beijing committing atrocities against the Falun Gong movement?', www.weeklystandard.com, 5 August 2006.
53. Healy, *Last Best Gifts*.
54. Goodwin, *Black Markets*.
55. John Moore, quoted in Andrews and Nelkin, *Body Bazaar*, p. 1.

Chapter two

1. Cited in Rebecca Skloot, 'Henrietta's Dance' (2000) *Johns Hopkins Magazine*, April, www.jhu.edu/~jhumag/0400web/01.html, accessed 27 March 2007.
2. H.A. Washington, 'Henrietta Lacks—an unsung hero', *Emerge*, October 1994, p. 29, quoted in Lori B. Andrews, 'Harnessing the benefits of biobanks' (2005) 22 *Journal of Law, Medicine and Ethics*, 22–30, at p. 25.
3. Anne Fagot-Largeault, 'Ownership of the human body: judicial and legislative responses in France', in Henk ten Have and Jos Welie (eds), *Ownership of the Human Body: Philosophical Considerations on the Use of the Human Body and Its Parts in Healthcare* (Dordrecht: Kluwer, 1998), pp. 115–40, p. 130.
4. Moore *v.* Regents of the University of California, 51 Cal 3rd 120, 793 P2d 479, 271 Cal Rptr 146 (1990), Cert. Denied 111 SCt 1388. My account of the Moore case is compiled from a variety of sources, including: James Boyle, *Shamans, Software and Spleens: Law and the Construction of the Information Society* (Cambridge, MA: Harvard University Press, 1996), pp. 21–4, 97–107; Lori Andrews and Dorothy Nelkin, *Body Bazaar: The Market for Human Tissue in the Biotechnology Age* (New York: Crown Publishers, 2001), pp. 1–2, 27–31; Rebecca Skloot, 'Taking the least of you', *New York Times*, 16 April 2006, accessed online 24 April 2006 at www.nytimes.com/magazine; Roger Brownsword, 'Biobank governance—business as usual?', paper presented at the fourth international PropEur workshop (Property Regulation in European Science, Ethics and Law), Tuebingen, Germany, 20–1 January 2005; Nils Hoppe, 'The curse of intangibility: tracing equitable IP rights in human biological material, or *Moore* revisited', in Christian Lenk, Nils Hoppe and Roberto Andorno (eds), *Ethics and Law of Intellectual Property: Current Problems in Politics, Science and Technology* (Aldershot: Ashgate, 2006), Chapter Ten; Catherine Waldby and Robert Mitchell, *Tissue Economies: Blood, Organs and Cell Lines in Late Capitalism* (Durham, NC: Duke University Press, 2006), Chapter Three; Loane Skene, 'Ownership of human tissue and the law' (2002) 3 *Nature Reviews Genetics*, 145–8; Charlotte H. Harrison, 'Neither Moore nor the market: alternative models for compensating contributors of human tissue' (2002) 28 *American Journal of Law and Medicine*, 77; Graeme Laurie, '(Intellectual) property? Let's think about staking a claim to our own genetic samples' (Edinburgh: AHRB Research Centre, 2004); and Baruch Brody, 'Intellectual property and biotechnology—the US internal experience, part 1' (2006) 16 *Kennedy Institute of Ethics Journal*, 1.
5. John Moore, interview in *Discover* magazine, quoted in Skloot, 'Taking the least of you', p. 4.

6. Harrison, 'Neither Moore nor the market'.
7. 6 CLR 406 [1908].
8. 'No party has cited a decision supporting Moore's argument that diseased human cells are "a species of tangible personal property that is capable of being converted".' (479 271 Cal Rptr 146, quoted in Waldby and Mitchell, *Tissue Economies*, p. 94).
9. Rohan Hardcastle, *Law and the Human Body: Property Rights, Ownership and Control* (Oxford: Hart, 2007).
10. Jean McHale, 'Waste, ownership and bodily products' (2000) 8 *Health Care Analysis*, 2, 123–35.
11. The classic case for English law is Sidaway *v.* Bethlem RHG [1985] 1 All ER 643. Lord Scarman, in his dissent, did propose a 'reasonable patient' standard, but the majority espoused the 'reasonable doctor' view. For further discussion of this distinction, see Donna Dickenson, *Risk and Luck in Medical Ethics* (Cambridge: Polity, 2003), pp. 66–71.
12. Moore *v.* Regents of the University of California, 793 P2d 479 (1990), 484.
13. Andrews and Nelkin, *Body Bazaar*, p. 31.
14. Moore *v.* Regents of the University of California, 793 P2d 479 (1990), 498.
15. Ibid., 506.
16. Brownsword, 'Biobank governance—business as usual?', p. 18.
17. Moore *v.* Regents of the University of California, 793 P2d 479 (1990), 517.
18. John Locke, *The Second Treatise on Civil Government* (1689), cited in G.A. Cohen, *Self-Ownership, Freedom and Equality* (Cambridge: Cambridge University Press, 1997) p. 209. See also Donna Dickenson, *Property, Women and Politics* (Cambridge: Polity, 1997), p. 78.
19. Justice George, in Moore *v.* University of California.
20. Lori Andrews, interviewed in Skloot, 'Taking the least of you', p. 5; Lori Andrews, 'Genes and patent policy: rethinking intellectual property rights' (2002) 3 *Nature Reviews Genetics*, 803–8.
21. Boyle, *Shamans, Software and Spleens*, p. 24.
22. Moore *v.* Regents of the University of California (1988), at 537.
23. Hoppe, 'The curse of intangibility'.
24. Greenberg et al. *v.* Miami Children's Hospital Research Institute, Inc., 208 F. Supp. 2d 918 (2002), 264 F Supp 2d 1064 (2003). My account of the case is taken from these sources among others: Peter Gorner, 'Parents suing over patenting of genetic test' (2000) *Chicago Tribune*, 19 November, http://home.iprimus.com.au, accessed 22 January 2007; John K. Borchadt, 'Children's parents sue over genetics patent' (2000) 1 *The Scientist*, 1122; Skloot, 'Taking the least of you'; David Evanier, 'Parents sue over Canavan test patent' (2001) www.jewishjournal.com,

accessed 22 January 2007; background information on *Greenberg v. Miami Children's Hospital et al.*, prepared by public interest (*pro bono*) attorneys in the case from Chicago Kent School of Law, including Prof. Lori Andrews (www.kentlaw.edu/classes/rstaudt/plustechlaw2003/canavan.htm, accessed 22 January 2007); and Donna M. Gitter, 'Ownership of human tissue: a proposal for Federal recognition of human research participants' property rights in their biological material' (2004) *Washington and Lee Review*, winter.

25. www.canavanfoundation.org/canavan:php, accessed 22 January 2007.
26. Quoted in Gorner, 'Parents suing over patenting of genetic test', p. 3.
27. Susan van Dusen, 'Whose genes are they?', online at Chicago Jewish Community/Jewish United Fund website, www.juf.org.news, accessed 22 January 2007.
28. M.R. Anderlik and M.A. Rothstein, 'Canavan decision favors researchers over families' (2003) 31 *Journal of Law and Medical Ethics*, 450–4.
29. Joint press release, 29 September 2003, www.canavanfoundation.org, accessed 22 January 2007.
30. R. *v* Kelly (1998) 3 All ER 741.
31. Quoted in Gorner, 'Parents suing over patenting of genetic test', p. 3.
32. Quoted in Lori Andrews, 'Shared patenting experiences: the roles of patients', paper presented at the fifth workshop of the EC PropEur project, Bilbao, December 2005, p. 5.
33. Matalon consistently denied that he had received royalties, however, claiming that it was a condition of his employment with Miami Children's Hospital that the institution should be allowed to patent the gene and retain all the proceeds.
34. Quoted in Evanier, 'Parents sue over Canavan test patent'.
35. Ibid.
36. Quoted in Gina Kolata, 'Who owns your genes?' (2000) *New York Times*, 15 May.
37. For example, P.J. van Diest and Julian Savulescu, 'For and against: no consent should be needed for using leftover body material for scientific purposes' (2002) 325 *British Medical Journal*, 648–51.
38. Professor Henry Greely, quoted in Kolata, 'Who owns your genes?', p. 6.
39. Mairi Levitt and Sue Weldon, 'A well-placed trust? Public perceptions of the governance of DNA databases' (2005) 15 *Critical Public Health*. 311–21.
40. London: George Allen and Unwin, 1970, second edition ed. A. Oakley and J. Ashton (London: LSE Books, 1997).
41. Waldby and Mitchell, *Tissue Economies.*
42. David B. Resnik, 'The commercialisation of human stem cells: ethical and policy issues' (2002) 10 *Health Care Analysis*, 127–54.

43. Goodwin, *Black Markets.*
44. Healy, *Last Best Gifts.*
45. *Counterfeit Money* (Chicago: University of Chicago Press, 1992).
46. John Frow, 'Gift and Commodity', in Frow (ed.), *Time and Commodity Culture: Essays in Cultural Theory and Postmodernity* (Oxford: Clarendon, 1997); Marcel Mauss, *The Gift: The Form and Reason for Exchange in Archaic Societies* (London: Routledge, 1990).
47. Soren Holm, 'Who should control the use of human embryonic stem cells? A defence of the donor's ability to control' (2006) 3 *Journal of Bioethical Inquiry*, 55–68.
48. Lesley A. Sharp, *Bodies, Commodities and Biotechnologies: Death, Mourning and Scientific Desire in the Realm of Human Organ Transfer* (New York: Columbia University Press, 2006).
49. Bartha M. Knoppers, 'Human genetic material: commodity or gift?' in R.F. Weir (ed.), *Stored Tissue Samples* (Iowa City: Iowa University Press, 1998), 226–35.

Chapter three

1. John Carvel, 'With love at Christmas: a set of stem cells' (2005) *Guardian*, 6 December, p. 7.
2. Interview for the programme 'Catalyst', Australian Broadcasting Service Television, 25 September 2004, quoted in Catherine Waldby and Robert Mitchell, *Tissue Economies: Blood, Organs and Cell Lines in Late Capitalism* (Durham and London: Duke University Press, 2006), p. 129.
3. G. Kogler et al., 'A new human somatic stem cell from placental cord blood with intrinsic pluripotent differentiation potential' (2004) 200 *Journal of Experimental Medicine*, 123–25.
4. E. Gluckman, H.A. Broxmeyer, A.D. Auerbach et al., 'Hematopoietic reconstitution in a patient with Fanconi's anemia by means of umbilical-cord blood from an HLA-identical sibling', *New England Journal of Medicine*, vol. 321 (1989), pp. 1174–8.
5. For example, the parents of Charlie Whittaker, who suffered from Diamond Blackfan anaemia, were permitted by the HFEA to use preimplantation genetic diagnosis to conceive a sibling who would be an exact tissue match for Charlie, since no one in the family was a suitable bone marrow donor. Charlie was given the 'all-clear' in 2005, demonstrating the efficacy of a cord blood transplant. Some commentators oppose the deliberate conception of a child as a means to the end of curing another child, however.
6. Jennifer Gunning, 'Umbilical cord blood banking: a surprisingly controversial issue', report for CCELS (Cardiff Centre for Ethics, Law and Science), 2005.

7. Nirmala Rai and Prathima Reddy, letter, 'Rapid Response' to Leroy C. Edozien, 'NHS maternity units should not encourage commercial banking of umbilical cord blood' (2006) 333 *British Medical Journal* 801–4, www.bmj.com/cgi/letters, accessed 31 January 2007.
8. Waldby and Mitchell, *Tissue Economies*, p. 125.
9. Cryo-Care advertising leaflet, p. 12.
10. Cryogenesis International website, accessed 2 January 2006.
11. Royal College of Obstetricians and Gynaecologists, 'Umbilical Cord Blood Banking', Scientific Advisory Committee Opinion Paper 2, originally issued in 2001, revised June 2006, p. 2.
12. The exception appears to be Cells4Life, which uses the method whereby blood is collected from the placenta after it has been expelled from the uterus. ('Response to the Royal College of Obstetricians and Gynaecologists' Scientific Advisory Committee Opinion Paper 2 on "Umbilical Cord Blood Banking"', 21 January 2007, www.cells4life.co.uk/news-details, accessed 31 January 2007.)
13. RCOG, 'Umbilical cord blood banking', p. 2.
14. P. Solves, R. Moraga, E. Saucedo et al., 'Comparison between two strategies for umbilical cord blood collection' (2003) 31 *Bone Marrow Transplant*, 269–73.
15. W.J. Prendiville, D. Elbourne and S McDonald, 'Active versus expectant management in the third stage of labour', *The Cochrane Database of Systematic Reviews* (2000) issue 3, Art. No. CD000007 (24 July). See also W. Prendiville and D. Elbourne (1989), 'Care during the third stage of labour', in I. Chalmers, M. Enkin and M.J.N.C. Keirse (eds), *Effective Care in Pregnancy and Childbirth* (Oxford: Oxford University Press, 1989), pp. 1145–69.
16. Cited in Saskia Tromp, *Seize the Day, Seize the Cord*, unpublished undergraduate medical dissertation, University of Maastricht (2001). My thanks to Saskia Tromp for making this citation known to me when I was co-supervising her dissertation.
17. Donna Dickenson and Paolo Vineis, 'Evidence-based medicine and quality of care', *Health Care Analysis*, vol. 10, no. 3 (2002), pp. 243–59, at p. 255.
18. Promotional literature from a variety of private cord blood banks, quoted in Royal College of Obstetricians and Gynaecologists, 'Umbilical Cord Blood Banking', Scientific Advisory Committee Opinion Paper 2, revised June 2006, p. 1.
19. *Pregnancy Weekly*, online at www.pregnancyweekly.com, accessed 31 January 2007.
20. Video clip online at www.pregnancyweekly.com, accessed 31 January 2007.
21. Patrick van Rheenen and Bernard J. Brabin, 'Late umbilical cord-clamping

as an intervention for reducing iron deficiency anaemia in term infants in developing and industrialised countries: a systematic review' (2004) 24 *Annals of Tropical Paediatrics*, 3–16.

22. H. Rabe, G. Reynolds and J. Diaz-Rossello, 'Early versus delayed umbilical cord clamping in preterm infants', *The Cochrane Database of Systematic Reviews*, 2004, issue 4, art. No. CD003248pub2, first published 18 October 2004, with a more recent review in volume 3, 17 May 2005. A review article by B. Lainez Villabona et al. ('Early or late umbilical cord clamping? A systematic review of the literature' [2005] 63 *Anales Pediatria*, 14–21) agrees that late clamping could diminish the proportion of children with low iron reserves at 3 months by 50 per cent, but notes that this study lost 40 per cent of patients during follow-up.
23. G.K. Hofmeyr, P.J.M Bex, R. Skapinker and T. Delahunt, 'Hasty clamping of the umbilical cord may initiate neonatal intraventricular hemorrhage' (1989) 29 *Medical Hypotheses*, 5. The validity of this study is disputed by Francesco Bartolini, Manuela Battaglia, Cinzia De Iulio and Girolano Sirchia, 'Response' (1995) 86 *Blood*, 12, 4900.
24. Norman Ende, 'Letter' (1995) 86 *Blood*, 12, Dec 15, 4699.
25. Cells4Life, 'Response to the Royal College of Obstetricians and Gynaecoloists' Scientific Advisory Committee Opinion Paper 2 on "Umbilical cord blood banking".' 27 January 2007, p. 2, www.cells4life.co.uk/news-details, accessed 31 January 2007.
26. RCOG, 'Umbilical cord blood banking', p. 4, citing the following studies: S.B. Dunnett and A.E. Rosser, 'Cell therapy in Huntington's Disease' (2004) 1 *Neuro Rx*, 394–405; B.O. Kim, H. Tian, K. Prasongsukarn et al., 'Cell transplantation improves ventricular function after a myocardial infarction: a preclinical study of human unrestricted somatic stem cells in a porcine model' (2005) 112 *Circulation*, 196–204; J. Leor, E. Guetta, P. Chouraqui et al., 'Human umbilical cord blood cells: a new alternative for myocardial repair?' (2005) 7 *Cytotherapy*, 251–7; H.E. Broxmeyer, 'Biology of cord blood cells and future prospects for enhanced clinical benefit' (2005) 7 *Cytotherapy*, 209–18; K.S. Kang, K.W. Sim, O.H. Yuh et al., 'A 37-year-old spinal cord-injured female patient, transplanted of multipotent stem cells from human UC blood, with improved sensory perception and mobility, both functionally and morphologically: a case study' (2005) 7 *Cytotherapy*, 368–73. There have also been early reports of the use of cord blood to produce insulin ('Cardinals to hear about breakthrough in umbilical cord research', *Sunday Times of Ireland*, 9 July 2006) and to make a two-centimetre 'miniature artificial liver' (Heidi Nicholls, 'Cord blood used to make miniature artificial liver', *Bionews Today*, 11 November 2006, www.bionews.org.uk, accessed 30 January 2007).

27. George J. Annas, 'Waste and longing: the legal status of placental blood banking' (1999) 340 *New England Journal of Medicine*, 1521–4.
28. Gluckman et al., 'Hematopoietic reconstitution in a patient with Fanconi's anemia by means of umbilical-cord blood from an HLA-identical sibling'.
29. Vanderson Rocha et al., 'Graft-versus-host disease in children who have received a cord-blood or bone marrow transplant from an HLA-identical sibling' (2000) 342 *New England Journal of Medicine*, 25, 1846–54, found that as an alternative to bone marrow for haematopoietic stem-cell transplantation, umbilical cord blood from a tissue-matched sibling may lower risk of graft-versus-host disease (GHVD), in a study of 113 recipients of cord blood compared with 2052 recipients of bone marrow.
30. Juliet N. Barker and John E. Wagner, 'Umbilical-cord blood transplantation for the treatment of cancer' (2003) 3 *Nature Reviews Cancer*, 526–32, which reports results for blood cancers treated with umbilical cord blood-derived haematopoietic stem cells in several studies involving both child and adult patients, confirming the lower incidence of graft-versus-host disease.
31. J.L. Wiemels, G. Cazzaniga, M. Daniotti, O.B. Eden, G.M. Addison, G. Masera et al., 'Prenatal origin of acute lymphoblastic leukaemia in children' (1999) 352 *Lancet*, 1499–503.
32. RCOG, 'Umbilical cord blood', citing C.G. Brunstein and L.E. Wagner, 'Umbilical cord blood transplantation and banking' (2006) 57 *Annual Review of Medicine*, 6–9.
33. Leroy C, Edozien, 'NHS maternity units should not encourage commercial banking of umbilical cord blood' (2005) 333 *British Medical Journal*, 801–4.
34. Ian Sample, 'Branson launches shared stem cell bank', *Guardian*, 2 February 2007, p. 11.
35. Cryo-Care advertising leaflet, p. 3.
36. Kenneth J. Moise Jr., 'Umbilical cord stem cells' (2005) 106 *Obstetrics and Gynecology*, 1393–1407.
37. Annas, 'Waste and longing'.
38. Moise, 'Umbilical cord stem cells', at p. 1405.
39. Carvel, 'With love at Christmas'.
40. RCOG, 'Umbilical cord blood banking', p. 7.
41. Sheila Kitzinger, *The New Experience of Childbirth* (London: Orion, 2004).
42. Anna Smyth, 'A leap in the dark?', *Scotsman*, 2 February 2007, www.news.scotsman.com, accessed 5 February 2007.
43. Quoted in ibid.
44. '"Stem cell ferry" gets round EU rules', BioNews, 2 May 2006, www.bionews.org.uk, accessed 30 January 2007.

45. Quoted in ibid.
46. Quoted in Smyth, 'A leap in the dark'.

Chapter four

1. Hwang, Woo-Suk, et al., 'Patient-specific embryonic stem cells derived from human SCNT blastocysts' (2005) 308 *Science*, 1777–83. This paper built on an earlier publication by Hwang and his colleagues, 'Evidence of a pluripotent human stem cell line derived from a cloned blastocyst' (2004) 303 *Science*, 1669–74.
2. I have taken this description in part from William B. Hurlbut, Robert P. George and Markus Grompe, 'Seeking consensus: a clarification and defense of altered nuclear transfer' (2006) 36 *Hastings Center Report*, 42–50.
3. Research by Daley and Jaenisch, summarised in Stephen S. Hall, 'Stem cells: a status report' (2006) 36 *Hastings Center Report*, 16–22.
4. Sarah Sexton, 'Transforming "waste" into "resource": from women's eggs to economics for women' (2005), paper presented at Reprokult workshop, Heinrich Boll Foundation, Berlin, 10 September. Sexton based her calculation on figures of 1.4 million diagnosed diabetics in the UK and an additional one million undiagnosed diabetics, creating a requirement of 2.4 million personal stem cell lines. (On the figures reported by Hwang, each cell line would require a donor to give about ten eggs.) There are approximately 5.8 million UK women aged between 20 and 34—the prime age for ova donation—implying that one in every two to three women would need to go through the egg retrieval process.
5. Personal communication from Paik Young-Gyung, 11 June 2007.
6. Ibid. My thanks to Paik Young-Gyung for clarifying this crucial role of Korean feminists themselves, which has largely been ignored in the media.
7. These figures are taken from the Opening Statement of Chairman Mark Souder, Congressional Subcommittee on Criminal Justice, Drug Policy and Human Resources, 'Human cloning and embryonic stem cell research after Seoul: examining exploitation, fraud and ethical problems in the research', 7 March 2006.
8. Phillan Joung, 'Breaking the silence—the aftermath of the egg and cloning scandal in South Korea' (2006), paper presented at the conference 'Connecting civil society—implementing basic values', Berlin, 17–19 March.
9. D. Zimmerman et al., 'Gender disparity in living renal transplant donation' (2000) 36 *American Journal of Kidney Diseases*, 534–40.
10. Clive Cookson, 'The cloning connection: cloned tissues from stem cells might beat immune rejection', ScientificAmerican.com, 27 June 2005.

11. Mark Souder, Opening Statement.
12. Joung, 'Breaking the silence'.
13. Paik Young-Gyung, 'Beyond bioethics: the globalized reality of ova trafficking and the possibility of feminist intervention' (2006), paper presented at the International Forum on the Human Rights of Women and Biotechnology, Seoul, 21 September.
14. Gretchen Vogel, 'Ethical oocytes: available for a price' (2006) 313 *Science*, 5784, 155.
15. E. Heitman and M. Schlachtenhaufen, 'The differential effects of race, ethnicity and socio-economic status on infertiity treatment', in C.B. Cohen (ed.), *New Ways of Making Babies: The Case of Egg Donation* (Bloomington, Indiana: Indiana University Press, 1996), pp. 188–212, at p. 194.
16. Suzanne Holland, 'Beyond the embryo: a feminist reappraisal of the stem cell debate', in Suzanne Holland and L. Lebacqz (eds), *The Human Embryonic Stem Cell Debate: Science, Ethics and Public Policy* (Cambridge, MA: MIT Press, 2001), pp. 73–86; Donna Dickenson, 'Commodification of human tissue: implications for feminist and development Ethics' (2002) 2 *Developing World Bioethics*, 1, 55–63; Donna Dickenson, 'Commodification of human tissue: implications for feminist and development Ethics' (Portuguese translation) in Debora Diniz (ed.), *Bioética Feminista Contemporânea* (Brazilia: Editora Letras Livres, 2003); Donna Dickenson, 'Ethics Watch: The threatened trade in human ova' (2004) 5 *Nature Reviews Genetics*, 2, 86.
17. Paik Young-Gyung, 'Beyond bioethics', p. 2.
18. Paul Lauritzen, 'Stem cells, biotechnology and human rights: implications for a posthuman future' (2005) 35 *Hastings Center Report*, 2, 25–33, at 25.
19. Melinda Cooper, 'The unborn born again: neo-imperialism, the evangelical right and the culture of life' (2006) 17 *Postmodern Culture*, 1.
20. Walter Benjamin, 'Capitalism as religion', in Marcus Bullock and Michael W. Jennings (eds), *Selected Writings of Walter Benjamin* (Cambridge, MA: Harvard University Press, 1999), pp. 288–91, originally published 1921.
21. Mark Henderson, 'Cloning benefits oversold, says UK scientist ' (2006) *The Times*, 18 December.
22. Jonathan Kimmelman, Francoise Baylis and Kathleen Cranley Glass, 'Stem cell trials: lessons from gene therapy research' (2006) 36 *Hastings Center Report*, 1, 23–6.
23. Stephen S. Hall, 'Stem cells: a status report' (2006) 36 *Hastings Center Report*, 1, 16–22, at p. 21.
24. David Cyranski, 'No end in sight for stem cell odyssey' (2006) 439 *Nature*, 658–9.
25. Quoted in Cyranski, 'No end in sight', p. 659.

26. Quoted in Phyllida Brown, 'Do we even need eggs?' (2006) 439 *Nature*, 655–7, at p. 655.
27. The title of an article describing Yamanaka's research in *Bioedge*, no. 253, 13 June 2007, online at www.newsletter@australasianbioethics.org, accessed 14 June 2007.
28. Dr LoMedico of the US-based Juvenile Diabetes Research Foundation, quoted in Hall, 'Stem cells', p. 21.
29. Dr Grompe of the same foundation, quoted in Hall, 'Stem cells', p. 21.
30. Dr Arnold Kriegstein, quoted in Hall, 'Stem cells', p. 21.
31. Quoted in Hall, 'Stem cells', p. 22.
32. Quoted in ibid., p. 21.
33. Souder, Opening Statement; Wolfgang Lilige, 'The case for adult stem cell research' (2002) *21st Century Science and Technology Magazine*, winter, online at www.sciencetech.com, accessed 7 March 2007.
34. J. Norsigian (2005) 'Egg donation for IVF and stem cell research: Time to weigh the risks to women's health', in Boston Women's Health Collective (ed.), *Our Bodies Ourselves*, ch. 25, online at www.ourbodies ourselves/orgbook/companion.
35. Kay Lazar, 'Wonder drug for men alleged to cause harm in women' (1999) *Boston Herald*, 22 August.
36. Ibid.
37. Bonnie Steinbock, 'Payment for egg donation and surrogacy' (2004) 71 *Mount Sinai Journal of Medicine*, 255–65.
38. Diane Beeson, statement before US Congressional Subcommittee on Criminal Justice, Drug Policy and Human Resources, 7 March 2006, p.3.
39. 'Woman died after starting IVF treatment', *Richmond & Twickenham Times*, 20 April 2005, online at www.richmondandtwickenhamtimes.co.uk/mayor/other/display.var.589076.0.0.php
40. M.J. Steigenga et al., 'Evolutionary conserved structures as indicators of medical risk: increased incidence of cervical ribs after ovarian hyperstimulation in mice' (2006) 56 *Animal Biology*, 1, 63–8.
41. Norsigian, 'Egg donation for IVF', p. 2.
42. A. Delavigne and S. Rozenberg, 'Epidemiology and prevention of ovarian hyperstimulation syndrome (OHSS): A review' (2002) 8 *Human Reproduction Update*, 559–77.
43. M.V. Sauer, R.J. Paulson and R.A. Lobo, 'Rare occurrence of ovarian hyperstimulation syndrome in oocyte donors' (1996) 52 *International Journal of Gynecology and Obstetrics*, 259–62; M.V. Sauer, 'Defining the incidence of serious complications experienced by oocyte donors: a review of 1,000 cases' (2001) 184 *American Journal of Obstetrics and Gynecology*, 277–8.
44. An anonymous website respondent, quoted in Steinbock, 'Payment for egg donation', p. 258.

45. HFEA Statement on Donating Eggs for Research, 21 February 2007, www.hfea.gov.uk, accessed 22 February 2007.
46. G. Pennings and M. DeVroey, 'Subsidized IVF treatment and the effect on the number of egg sharers' (2006) 1 *Reproductive Medicine Online*, 8–10.
47. My thanks to Dr Neville Cobb of the University of Edinburgh for making these statistics available to me, drawn from his extensive analysis of HFEA inspection reports.
48. Donna Dickenson, quoted in James Randerson, 'IVF women will be paid to offer spare eggs for research', *Guardian*, 28 July 2006, p. 12.
49. HFEA Statement on Donating Eggs for Research, 21 February 2007, www.hfea.gov.uk, accessed 22 February 2007.
50. Esther Heijnen, Marinus Eijkemans, Cora de Klerk et al., 'A mild treatment strategy for in-virto fertilisation: a randomised non-inferiority trial' (2007) 369 *Lancet*, 743–9.
51. David Magnus and Mildred K. Cho, 'Issues in occyte donation for stem cell research' (2005) 308 *Science*, 1747–8.
52. Denis Campbell, 'Women will be paid to donate eggs for science' (2007) *Observer*, 18 February, p. 1.
53. Ethics Committee of the American Society for Reproductive Medicine, 'Financial incentives in recruitment of oocyte donors' (2004) 82 *Fertility and Sterility* supplement, S240–4.
54. Editorial, 'Safeguards for donors' (2006) 442 *Nature*, 7103, 601.
55. Insoo Hyun, 'Fair payment or undue inducement?' (2006) 442 *Nature*, 7103, 629–30, at p. 629.
56. Editoral, 'Safeguards for donors', p. 601.
57. M.D. Althuis et al. (2005) 161 *American Journal of Epidemiology*, 607–15, cited in Helen Pearson, 'Health effects of egg donation may take decades to emerge' (2006) 442 *Nature*, 7103, 607–8, at p. 8.
58. S.D. Halpern et al. (2004) 164 *Archives of Internal Medicine* 801–3, cited in Hyun, 'Fair payment or undue inducement?', p. 630.
59. Kieran Healy, *Last Best Gifts: Altruism and the Market for Human Blood and Organs* (Chicago and London: University of Chicago Press, 2006), p. 24.
60. Michele Goodwin, *Black Markets: The Supply and Demand of Body Parts* (Cambridge and New York: Cambridge University Press, 2006), p. 43.
61. Soren Holm, 'Who should control the use of human embryonic stem cell lines?: a defence of the donor's ability to control' (2006) 3 *Journal of Bioethical Inquiry*, 55–68.

Chapter five

1. Paul Oldham, 'The patenting of animal and plant genomes' (2006), paper presented at the seventh workshop of the European Commission

PropEur project (Property Regulation in European Science, Ethics and Law), Paris, 5–6 May.

2. Except in Gregory Maguire's novel *Wicked: The Life and Times of the Witch of the West* (New York: Harper Collins, 1995), where the witch is a doughty if ultimately unsuccessful freedom fighter against the Wizard, an infamous tyrant.
3. K. Jensen and F. Murray, 'International patenting: the landscape of the human genome' (2005) 310 *Science*, 239–40.
4. David L. Finegold et al., *Bioindustry Ethics* (Amsterdam: Elsevier, 2005), p. 203.
5. Finegold et al., *Bioindustry Ethics*, p. 202, stating the argument of Sciona's critics, the activist organisation Genewatch.
6. Lori B. Andrews, 'Genes and patent policy: Rethinking intellectual property rights' (2002) 3 *Nature Reviews Genetics*, 803–8.
7. Figures from www.breastcancer.org, accessed 16 March 2007.
8. 447 US 303.
9. Howard Florey/Relaxin, *European Patent Office Reports* (1995), p. 541. For a more complete discussion of the Relaxin case, see Derek Beyleveld and Roger Brownsword, 'Patenting human genes: legality, morality and human rights', in J.W. Harris (ed.), *Property Problems: From Genes to Pension Funds* (London: Kluwer Law International, 1997), pp. 9–24.
10. Pilar Ossorio, 'Legal and ethical issues in biotechnology patenting', in J. Burley and J. Harris, *A Companion to Genethics* (Oxford: Blackwell, 2002), pp. 408–19.
11. Bronwyn Parry, *Trading the Genome: Investigating the Commodification of Bio-Information* (New York: Columbia University Press, 2004), p. 50.
12. S.J.R. Bostyn, 'One patent a day keeps the doctor away? Patenting human genetic information and health care' (2004) 7 *European Journal of Health Law*, 229–64.
13. Lori Andrews, 'Shared patenting experiences: the role of patients' (2005), paper presented at workshop of the European Commission project PropEur, Bilbao, 6 December. Patent claim available at blather.newdream.net/p/patent.html.
14. Mark M. Hanson, 'Religious voices in biotechnology: the case of gene patenting' (1997) 27 *Hastings Center Report*, no. 6, special supplement.
15. Eisenberg, 'How can you patent genes?'
16. Diamond *v.* Diehr, 450 U.S.175, 185 (1981).
17. Eisenberg, 'How can you patent genes?', p. 4.
18. Article 27 of TRIPS.
19. Nuffield Council on Bioethics, *The Ethics of Patenting DNA* (London: Nuffield Council, 2002), p. 29.
20. Diamond *v.* Chakrabarty, 447 U.S. 303 (1980).

21. Funk Brothers Seed Co. *v.* Kalo Inoculant Co., 333 US 127 (1948), cited in Parry, *Trading the Genome*, p. 85, fn. 52.
22. Diamond *v.* Chakrabarty, at p. 309.
23. 333 US 118 (1947).
24. E. Richard Gold, *Body Parts: Property Rights and the Ownership of Human Biological Materials* (Washington, DC: Georgetown University Press, 1996), p. 81.
25. This latter point is made by Gold, in *Body Parts*, pp. 83–4.
26. Gold, *Body Parts*, pp. 84–5.
27. Diamond *v.* Chakrabarty, p. 307, citing Kewanee Oil Co. *v.* Bicton Co., 416 US 470, 480 (1974).
28. Diamond *v.* Chakrabarty, p. 320 (original emphasis).
29. Parry, *Trading the Genome*, p. 87.
30. Nuffield Council, *The Ethics of Patenting DNA*, p. 49.
31. Bostyn, 'One patent a day', p. 233.
32. Ibid., p. 234.
33. Parry, *Trading the Genome*, p. 88.
34. Jordan Paradise, Lori Andrews and Timothy Caulfield, 'Patents on human genes: an analysis of scope and claims' (2006) 307 *Science*, 1566–7.
35. Melinda Cooper, 'The unborn born again: neo-liberalism, the evangelical right and the culture of life' (2006) 17 *Postmodern Culture*, 1.
36. Cited in Parry, *Trading the Genome*, p. 93.
37. Article 6 of the 1998 EC Directive further limits the *ordre public* exclusion by invoking an extreme utilitarian argument: provided some public benefit is likely to result from exploitation of the patent, the exclusion is unlikely to be enforced. See W.R. Cornish, M. Llewelyn and M. Adcock, *Intellectual Property Rights (IPRs) and Genetics: A Study into the Impact and Management of Intellectual Property Rights within the Healthcare Sector* (Cambridge: Cambridge Genetic Knowledge Park, July 2003), section 2.C.3 (b), 'Morality'). Although the notion of *ordre public* is confined to European patent law, the US Patent Law 2000 excludes inventions whose use is inherently immoral.
38. Sigrid Stercx and Julian Cockbain, 'Stem cell patents and morality: the European Patent Office's emerging policy', forthcoming.
39. Lori Andrews and Dorothy Nelkin, *Body Bazaar: The Market for Human Tissue in the Biotechnology Age* (New York: Crown, 2001), p. 50.
40. I base my narrative on an account by the Director of the Tonga Human Rights and Democracy Movement, Lopeti Senituli: L. Senituli, 'They came for sandalwood, now the b...s are after our genes!', paper presented at the conference 'Research ethics, tikanga Maori/indigenous and protocols for working with communities'. Wellington, New Zealand: 10–12 June 2004.

41. Senituli, 'They came for sandalwood', p. 3.
42. Andrews and Nelkin, *Body Bazaar*, p. 79, in a chapter discussing instances from Tristan da Cunha, the Human Genome Diversity Project, the Hagahai of Papua New Guinea and other similar examples to the Tongan one. See also an anonymous article in *Nature* (18 November 2004) describing a parallel attitude among the indigenous peoples of Vancouver Island, who donated blood for research into the genetic causes of rheumatoid arthritis, a disease that is rampant in their tribe. Twenty years later they were incensed to discover that the specimens have been used for other research—including a project on the sensitive issue of the spread of lymphotropic viruses by intravenous drug abuse. Leaders of the Nuu-chah-nulth (Nootka) tribe described the research as another example of exploitation of indigenous peoples and demanded the return of the samples.
43. Dorothy Nelkin and Susan Lindee, *The DNA Mystique: The Gene as a Cultural Icon* (New York: W.H. Freeman and Co., 1995), p. 39.
44. Senituli, 'They came for sandalwood', p. 4.
45. H.M. Mead, *Tikanga Maori: Living by Maori Values* (Wellington, New Zealand: Huia Publishers, 2003), p. 45.
46. Senituli, 'They came for sandalwood', p. 3.
47. Ibid., p. 4.
48. For example, John Harris ('An ethically defensible market in human organs', with C. Erin [2002] 325 *British Medical Journal*, 114–15), Julian Savulescu ('Is the sale of body parts wrong?' [2003] 16 *Journal of Medical Ethics*, 117–19); and Janet Radcliffe Richards ('The case for allowing kidney sales' [1998] 352 *Lancet*, 1950–2).
49. Alain Claeys, *Rapport sur les conséquences des modes d'appropriation du vivant sur les plans économique, juridique et éthique*, Office Parlementaire d'Évaluation des Choix Scientifiques et Technologiques, March 2004, http://www.assemblee-nationale.fr/12/rap-eocst/i1487.asp.
50. Timothy Caulfield, E. Richard Gold and Mildred K. Cho, 'Patenting human genetic material: refocusing the debate' (2000) 1 *Nature Reviews Genetics*, 227–31.
51. Dominique Memmi, *Les gardiens du corps: dix ans de magistère bioéthique* (Paris: Éditions de l'École des Hautes Études en Sciences Sociales, 1996).
52. Although the government never formally adopted the view that 'French DNA' was part of such a thing as 'the genetic patrimony of the nation', that position was taken publicly by health minister Jean-François Mattei.
53. Paul Rabinow, *French DNA: Trouble in Purgatory* (Chicago: University of Chicago Press, 1999), p. 126.
54. Comité Consultatif National d'Éthique, *Recherche biomédicale et respect de la personne humaine* (Paris: Documents Français), paragraph 2.3.2.

('Un irrespect intolérable de la personne, une violation radicale de notre droit, une menace de pourrissement pour toute notre civilisation'.)

55. Lucien Sève, *Qu'est-ce que la personne humaine? Bioéthique et démocratie* (Paris: La Dispute, 2006), p. 21, translation mine.
56. Ibid., p. 22, translation mine.
57. Comité Consultatif National d'Éthique, *Opinion No. 1: On sampling of dead human embryonic and foetal tissue for therapeutic, diagnostic and scientific purposes* (May 22, 1984), www.ccne-ethique.fr/english/avis/a_001.htm.
58. CCNE *Opinion no. 27,* www.ccne-ethique.fr/english/avis/a_027.htm, p. 2.
59. Stated in articles 16 and 1128 of the Civil Code.
60. CCNE *Opinion no. 21*, www.ccne-ethique.fr/english/avis/a_021.htm, p. 2.
61. For the effect of this emphasis in research ethics, see Giovanni Maio, 'The cultural specificity of research ethics—or why ethical debate in France is different' (2002) 28 *Journal of Medical Ethics*, 147–50.
62. CCNE opinion no. 74, 'Umbilical cord blood banks for autologous use or research', 12 December 2002, www.ccne-ethique.fr/english/avis/a_74.htm.
63. Marie-Angèle Hermitte, *Le sang et le droit: essai sur la transfusion sanguine* (Paris: Éditions du Seuil, 1996), p. 15.
64. Seve, *Qu'est-ce que la personne humaine?*, p. 25.
65. CCNE Opinion no. 93, 'Commmercialisation des cellules souches humaines et autres lignées cellulaires' (2006), www.ccne-ethique.fr/francais/avis/a_093.htm, accessed 12 April 2007.
66. CCNE opinion no. 93, section 5b.
67. The bioethics law of 6 August 2004 (law no. 2004–800) allows research on embryonic and foetal stem cell lines, under authorisation from the Biomedicine Agency, provided that the embryos were conceived *in vitro* through IVF and are no longer required by the commissioning couple. Embryos may not be created specifically for research purposes.
68. CCNE opinion no. 93, section 2c.
69. 'However, we must not let ourselves be deluded into thinking that morality is all on the side of the public, and immorality on the side of the private' (section II, translation mine).
70. Agence de la Biomédecine, *Quelles revisions de la loi de bioéthique?*, proceedings of a session convened 7 February 2007, available from www.agence-biomedecine.fr, accessed 21 May 2007.
71. In the above proceedings, p. 53.
72. Blandine Grosjean, 'Ces femmes infertiles contraintes à l'éxil', *Libération*, 20 August 2004.
73. A possible exception should be made for the surprise decision in which

US authorities revoked three key patents on human embryonic stem cells held by the University of Wisconsin (Erica Cheek, 'Patenting the obvious?' (2007) *Nature*, 3 May, online at www.nature.com, accessed 9 May 2007).

Chapter six

1. RAND corporation report, summarised in Rebecca Skloot, 'Taking the least of you: the tissue-industrial complex' (2006) *New York Times*, 16 April, available at www.nytimes.com/2006/04/16/magazine/16tissuehtml, accessed 24 April 2006.
2. For a summary of the ethical issues in the Alder Hey and Bristol scandals, see Veronica English, Rebecca Mussell, Julian Sheather and Ann Sommerville, 'Ethics briefings: retention and use of human tissue' (2004) 30 *Journal of Medical Ethics*, 235–6.
3. Russell Jenkins, 'Families demand investigation to end Sellafield's culture of silence' (2007), *The Times*, 18 April.
4. Christine Buckley, 'Body parts taken from workers at more nuclear sites' (2007) *The Times*, 20 April.
5. Skloot, 'Taking the least of you: the tissue-industrial complex'.
6. http://prostatsecure.wustl.edu, website of the School of Medicine of Washington University in St Louis, accessed 27 March 2007.
7. Quoted in Lori Andrews, 'Who owns your body? A patient's perspective on *Washington University v. Catalona*' (2005) 34 *Journal of Law, Medicine and Ethics*, 398–9.
8. *Amicus curiae* brief by Lori Andrews and Julie Burger on behalf of the People's Medical Society, No. 06-2288 and No. 06-2301, US Court of Appeals for the Eighth Circuit, Washington University *v.* William J. Catalona and Richard Ward et al., 21 July 2006, p. 10. My thanks to Lori Andrews and Julie Burger for very kindly making this brief available to me.
9. Patients' rights expert Prof. Ellen Wright-Clayton, quoted in Andrews, 'Who owns your body?', p. 401.
10. Richard Ward, a long-standing patient of Dr Catalona, quoted ibid.
11. James Ellis, another patient of Dr Catalona, quoted ibid.
12. Quoted in Rebecca Skloot, 'Tissue ownership update, II: more on Catalona', www.rebeccaskloot.blog.spot.com, 18 April 2006, accessed 27 March 2007.
13. Washington University *v.* William J. Catalona, 437 F Supp 2d, ESCD Ed Mo 2006.
14. Ibid.
15. Ibid.
16. Andrews, 'Who owns your body?', p. 401.

17. Mark A. Rothstein, 'Expanding the ethical analysis of biobanks' (2005) 33 *Journal of Law, Medicine and Ethics*, 1, 89–101.
18. In Missouri state law, including Estate of Bean *v.* Hazel, 972 S.W.2d 290, 293 (Mo. 1998), cited in *amicus curiae* brief filed by Andrews and Burger.
19. Rebecca Skloot, 'Big news on the tissue research front: a congressional investigation into researchers profiting off tissues without consent', 'Culture Dish' weblog, 14 June 2006, www.rebeccaskloot.blogspot.com, accessed 27 March 2007.
20. Stock *v.* Augsburg College, Minn. Ct. App. 2002, cited in Andrews, 'Who owns your body?', fn. 52, p. 406.
21. Rohan Hardcastle, *Law and the Human Body: Property Rights, Ownership and Control* (Oxford: Hart Publishing, 2007), p. 23.
22. Andrews, 'Who owns your body?', p. 402.
23. Karen Gottlieb, 'Human biological samples and the law of property: the trust as a model for biological repositories', in R.F. Weir (ed.), *Stored Tissue Samples: Ethical, Legal and Public Policy Implications* (Iowa City: Iowa University Press, 1998), pp. 183–97.
24. David E. and Richard N. Winickoff, 'The charitable trust as a model for genomic biobanks' (2003) 349 *New England Journal of Medicine*, 12, 1180–4. For a commentary on the original paper by the Winickoffs, see J. Otten, H. Wyle and G. Phelps, 'The charitable trust as a model for genomic banks' (2004) 350 *New England Journal of Medicine*, 85–6. On the use of the model by the VA, see David E. Winickoff and Larissa B. Neumann, 'Towards a social contract for genomics: property and the public in the "biotrust" model' (2005) 1 *Genomics, Society and Policy*, 3, 8–21, footnote 4.
25. Catherine Waldby and Robert Mitchell, *Tissue Economies: Blood, Organs and Cell Lines in Late Capitalism* (Durham, NC: Duke University Press, 2006), p. 79. For example, UK Biobank's literature often describes the bank as the 'steward' of the samples it contains.
26. See, for example, Hilary Rose, 'An ethical dilemma: the rise and fall of UmanGenomics—the model biotech company?' (2004) 425 *Nature*, 123–4.
27. Winickoff and Neumann, 'Towards a social contract', 11.
28. Roger Brownsword, 'Biobank governance: business as usual?' (2005), paper presented at the fourth workshop of the EC PropEur project, Tuebingen, 21 January, p. 38.
29. Andrews and Burger, *amicus curiae* brief for People's Medical Society, p. 4.
30. Ibid., p. 5.
31. Ibid., p. 5, fn. 2.
32. Naomi Klein, *No Logo* (London: Picador, 2000).
33. Andrews and Burger, *amicus curiae* brief, p. 7.

34. Ibid., pp. 11–12, numbering altered.
35. Committee on Human Genome Diversity, Commission on Life Sciences, US National Research Council, *Evaluating Human Genetic Diversity* (National Academy Press, 1997), p. 65, quoted in Andrews and Burger, *amicus curiae* brief, p. 14.
36. Andrews and Burger, *amicus curiae* brief, p. 29.
37. No. 06-2286, Washington University *v.* William J. Catalona and No. 06-2301, Washington University *v.* Richard N. Ward et al., p. 14 (490 F 3d 667, 8th Cir 2007).
38. No. 06-2286, Washington University *v.* William J. Catalona and No. 06-2301, Washington University *v.* Richard N. Ward et al., p. 9.

Chapter seven

1. Lorelle Phillips, interviewed in Ralph Gardner Jr., 'Looks to die for' *New York* magazine, www.nymag.com, accessed 9 April 2007.
2. Ibid.
3. Alex Kuczynski, *Beauty Junkies: Inside Our $15 Billion Obsession with Cosmetic Surgery* (New York: Doubleday, 2006), p. 122.
4. Kuczynski, *Beauty Junkies*, p. 203.
5. Ibid., p. 268.
6. Quoted in Kuczynski, *Beauty Junkies*, p. 67.
7. Kuczynski, *Beauty Junkies*, p. 29.
8. 'Medical tourism soaring' (2007) *Bioedge*, 6 June, online at www.australiabioethics.org, accessed 7 June 2007.
9. The clever title of a history of cosmetic surgery by Elizabeth Haiken (Baltimore, MD: Johns Hopkins University Press, 1997).
10. Kuczynski, *Beauty Junkies*, p. 155.
11. Ibid., p. 8.
12. For example, Sander L. Gilman, *Making the Body Beautiful: A Cultural History of Aesthetic Surgery* (Princeton, NJ: Princeton University Press, 1999).
13. Kuczynski, *Beauty Junkies*, p. 11.
14. Haiken, *Venus Envy*, p. 4.
15. Virginia L. Blum, *Flesh Wounds: The Culture of Cosmetic Surgery* (Berkeley, CA: University of California Press, 2003), p. 5.
16. Subscriber to website www.BeautyAddiction.com, quoted in Kuczynski, *Beauty Junkies*, p. 82.
17. Gilman, *Making the Body Beautiful*, p. 306.
18. Kuczynski, *Beauty Junkies*, p. 112.
19. Carolyn Latteier, *Breasts: The Woman's Perspective on an American Obsession* (New York: Haworth Press, 1998), p. 4, quoted in Kuczynski, *Beauty Junkies*, p. 249.

20. Gilman, *Making the Body Beautiful*, p. 304.
21. Between the two French cases, another partial face transplant was carried out in China on a man who had been mauled by a bear. Little detail has emerged about this case, however.
22. My thanks to Professor Laurent Lantieri for this quotation and his enormously helpful description of the full clinical details of the procedure he performed, as summarised in a seminar at the *École de Médecine*, Paris-V, Site Cochin, 30 May 2007, and to Dr Simone Bateman of the *Centre National de la Recherche Scientifique* (CNRS) for organising the seminar and inviting me to be a co-presenter.
23. Jean-Michel Bader, 'La greffe de visage s'attaque à une grave déformation' ('Face transplants tackle a serious deformity'), *Le Figaro*, 25 January 2007, www.lefigaro.fr/science, accessed 17 April 2007.
24. Michael Freeman and Pauline Abou Jaoude, 'Justifying surgery's last taboo: the ethics of face transplants' (2007) 33 *Journal of Medical Ethics*, 76–81, at p. 80.
25. Ibid., p. 78.
26. National Health Service UK Transplant Service, *UK Transplant Activity Report 2003–2004*, cited in Richard Huxtable and Julie Woodley, 'Gaining face or losing face? Framing the debate on face transplants' (2005) 19 *Bioethics*, 505–22, at p. 512.
27. Royal College of Surgeons Working Party, *Facial Transplantation* (London: Royal College of Surgeons, 2006, second edition), p. 11.
28. Ibid., p. 7.
29. O.P. Wiggins, J.H. Barker, S. Martinez et al., 'On the ethics of facial transplantation' (2004) 4 *American Journal of Bioethics*, 1–12. For a more complete treatment of the risks of failure, see G. Agich and M. Siemionov, 'Until they have faces: the ethics of facial allograft transplantation' (2005) 31 *Journal of Medical Ethics*, 707–9, and the response by Richard Huxtable and Julie Woodley, '(When) will they have faces? A response to Agich and Siemionov' (2006) 32 *Journal of Medical Ethics*, 403–4.
30. Royal College of Surgeons Working Party, *Facial Transplantation*, p. 11.
31. Huxtable and Woodley, 'Gaining face or losing face?', p. 513.
32. Ibid.
33. Michael J. Brenner et al., 'The spectrum of complications of immunosuppression: is the time right for hand transplantation?' (2002) 84 *Journal of Bone and Joint Surgery*, 1861–70, p. 1865.
34. Paragraph 32, cited in Royal College of Surgeons Working Party, *Facial Transplantation*, p. 32.
35. ('Je suis revenue sur la planète des humains'), interview in *Le Monde*, 7 July 2007.
36. Huxtable and Woodley, 'Gaining face or losing face?', p. 509, describing

an operation performed by Professor Giuseppe Spriano, reported on 2 February 2003.

37. The question of payment is a vexed one. In the case of the first human hand transplant, detailed in the next section, emphasis was laid in the proposal to the London hospital ethics committee on the patient's ability to pay privately for his immunosuppressants. Except for Freeman and Jaoude, however, few commentators have asked whether a lifelong regime of immunosupressants (for what is after all elective surgery) can be justified, when other major cuts in health service spending are being made.
38. Huxtable and Woodley, 'Gaining face or losing face?', p. 514.
39. M.A. Sanner, 'People's feelings and ideas about receiving transplants of different origins: questions of life and death, identity and nature's border' (2001) 15 *Clinical Transplant*, 19–27; J. Craven and G.M. Rodin, *Psychiatric Aspects of Organ Transplantation* (New York: Oxford University Press, 1992).
40. The legal case *Re M* (1999), cited in Huxtable and Woodley, 'Gaining face or losing face?', p. 516.
41. Kuczynski, *Beauty Junkies*, p. 65.
42. Huxtable and Woodley, '(When) will they have faces?', p. 404.
43. S. Katz and S. Kravetz, 'Facial plastic surgery for people with Down syndrome: research findings and their professional and social implications' (1989) 94 *American Journal of Mental Retardation*, 101–10; A.W. Frank, 'Emily's scars: surgical shapings' (2004) 34 *Hastings Center Report*, 18–29, both cited in Freeman and Jaoude, 'Justifying surgery's last taboo', footnotes 58 and 59.
44. Royal College of Surgeons Working Party, *Facial Transplantation*, p. 26.
45. Ibid., p. 27 and p. 29.
46. Donna Dickenson and Guy Widdershoven, 'Ethical issues in limb transplants' (2001) 15 *Bioethics*, 110–24, at p. 122.
47. Peter A. Clark, 'Face transplantation: Part II—an ethical perspective' (2005) 11 *Medical Science Monitor*, 41–7, at p. 46.
48. Freeman and Jaoude, 'Justifying surgery's last taboo', p. 79
49. Royal College of Surgeons Working Party, *Facial Transplantation*, p. 3.
50. Marco Lanzetta et al., 'The international registry in hand and composite tissue transplantation' (2005) 79 *Transplantation*, 1210–14.
51. N.F. Jones, 'Concerns about human hand transplantation in the 21st century' (2002) 27 *Journal of Hand Surgery*, 771–87, at p. 771.
52. Lanzetta et al., 'The international registry on hand and composite tissue transplantation'.
53. Brenner et al., 'The spectrum of complications of immunosuppression: is the time right for hand transplantation?', p. 1861.
54. Dickenson and Widdershoven, 'Ethical issues in limb transplants'.

55. M. Shaw (ed.), *After Barney Clark: Reflections on the Utah Artificial Heart Program* (Austin, Texas: University of Texas Press, 1984).
56. Carl Elliott, 'Doing harm, living organ donors, clinical research and *The Tenth Man*' (1995) 21 *Journal of Medical Ethics*, 91–6.
57. Martin M. Klapheke et al., 'Psychiatric assessment of candidates for hand transplantation' (2000) 20 *Microsurgery*, 453–7.
58. J.M. Dubernard et al., 'Human hand allograft: report on the first six months' (1999) 353 *Lancet*, 1315–20.
59. C. Hallam, quoted in *International Herald Tribune*, 21 October 2000.
60. Craven and Rodin, *Psychiatric Aspects of Organ Transplantation*, pp. 169–71.
61. Raymond Tallis, *The Hand: A Philosophical Inquiry into Human Being* (Edinburgh: Edinburgh University Press, 2003), p. 9.
62. Quoted in Francoise Baylis, 'A face is not just like a hand: *pace* Barker' (2004) 4 *American Journal of Bioethics*, 30–2, at p. 32.
63. Ibid.

Chapter eight

1. Hervé Juvin, *L'avènement du corps* (Paris: Le Débat, Gallimard, 2005).
2. Daniel Bell, *The End of Ideology: On the Exhaustion of Political Ideas in the Fifties* (New York: The Free Press, 1960).
3. Dan McDougall, 'Wives fall prey to kidney trade' (2007) *Observer*, 18 February, p. 43.
4. British Medical Association, *The Medical Profession and Human Rights: Handbook for a Changing Agenda* (London: Zed Books, 2001), Chapter Eight, 'Trade in Organs', pp. 193–204.
5. David Matas and David Kilgour, *Bloody Harvest: Revised Report into Allegations of Organ Harvesting from Falun Gong Prisoners in China*, 31 January 2007, p. 27, downloaded from www.investigation.go. saveinter.net, 14 May 2007. My thanks to Dr Emiko Tarasaki for alerting me to this report.
6. Matas and Kilgour, *Bloody Harvest*, p. 2.
7. China International Transplantation Network Assistance Centre website, cited in Matas and Kilgour, *Bloody Harvest*, p. 16.
8. Matas and Kilgour, *Bloody Harvest*, p. 27.
9. James Boyle, 'Fencing off ideas: enclosure and the disappearance of the public domain', Interactivist Info Exchange, available at http://slash.autonomedia.org.analysis, p. 5, accessed 10 September 2004, version of 'The second enclosure movement and the construction of the public domain' (2003) 66 *Law and Contemporary Problems*, 33–74.

10. Comité Consultatif National d'Éthique (CCNE), Opinion No. 74, *Umbilical Cord Blood Banks for Autologous Use or Research* (Paris: CCNE, 2002).
11. James Boyle, *Shamans, Software and Spleens: Law and the Construction of the Information Society* (Cambridge, MA: Harvard University Press, 1996).
12. Boyle, 'Fencing off ideas', p. 4.
13. Garrett Hardin, 'The tragedy of the commons' (1968) 162 *Science*, 1243.
14. Mark Heller and Rebecca Eisenberg, 'Can patents deter innovation? The anticommons in biomedical research' (1998) 280 *Science*, 698–701.
15. The sailor John MacCulloch, describing sailing past the townships of Upper and Lower Hallaig early in the nineteenth century, quoted in Roger Hutchinson, *Calum's Road* (Edinburgh: Birlinn, 2006), p. 9.
16. Brian Goldman, 'HER2: the patent "genee" is out of the bottle' (2007) 176 *Journal of the Canadian Medical Association*, 1443–4.
17. Boyle, *Shamans, Software and Spleens*, p. 23.
18. Donna Dickenson, 'Commodification of human tissue: implications for feminist and development ethics' (2002) 2 *Developing World Bioethics*, 55–63.
19. Boyle, *Shamans, Software and Spleens*, p. 24.
20. Ibid., p. 177.
21. Ibid., p. 128.
22. For further detail, see Donna Dickenson, *Property, Women and Politics* (Cambridge: Polity, 1997), Chapter Three.
23. Donna Dickenson, 'The lady vanishes: what's missing from the stem cell debate' (2006) 3 *Journal of Bioethical Inquiry*, 43–54.
24. For example, when the UK's influential Nuffield Council on Bioethics issued its report *Stem Cell Therapies: The Ethical Issues* in 2000, the ethical debate turned solely on the question of whether removal of stem cells constituted disrespect for the embryo. Concluding that it did not, the Council then saw no further reasons why stem cell research shouldn't proceed full steam ahead. No consideration whatsoever was given to the question of whether somatic cell nuclear transfer methods in stem cell research might involve 'disrespect' for or exploitation of women donating the necessary eggs.
25. For a more extended justification of this argument, see Donna Dickenson, *Property in the Body: Feminist Perspectives* (Cambridge: Cambridge University Press, 2007).
26. Walter Laqueur, *Making Sex: Body and Gender from the Greeks to Freud* (Cambridge, MA: Harvard University Press, 1992); Matthew Cobb, *The Egg and Sperm Race: The Seventeenth-Century Scientists Who Unravelled the Secrets of Sex, Life and Growth* (London: Free Press, 2006).

27. Tom Parfit, 'Beauty salons fuel trade in aborted babies' (2005), *Guardian Unlimited*, 17 April, online at www.guardian.co.uk, accessed 20 April 2005.
28. Dennis Normile and Charles Mann, 'Asia jockeys for stem cell lead' (2005) 307 *Science*, 660–4, cited in Emily Jackson, 'Fraudulent stem cell research and respect for the embryo' (2006) 1 *Biosocieties*, 349–6, at p. 356.
29. See Dickenson, *Property, Women and Politics*, for further details.
30. Maurice Merleau-Ponty, *The Visible and the Invisible* (Evanston, IL: Northwestern University Press, 1968), p. 137.

Bibliography

Agence de la Biomédecine (France), *Quelles revisions de la loi de bioéthique?*, proceedings of a session convened 7 February 2007, available from www.agence-biomedecine.fr, accessed 21 May 2007.

Agich, G., and Siemionov, M., 'Until they have faces: the ethics of facial allograft transplantation' (2005) 31 *Journal of Medical Ethics*, 707–9.

Anderlik, M.R., and Rothstein, M.A., 'Canavan decision favors researchers over families' (2003) 31 *Journal of Law and Medical Ethics*, 450–4.

Andrews, Lori B., 'Genes and patent policy: rethinking intellectual property rights' (2002) 3 *Nature Reviews Genetics*, 803–8.

—— 'Harnessing the benefits of biobanks' (2005) 22 *Journal of Law, Medicine and Ethics*, 22–30.

—— 'Shared patenting experiences: the role of patients' (2005), paper presented at workshop of the European Commission project PropEur, Bilbao, 6 December.

—— 'Who owns your body? A patient's perspective on *Washington University v. Catalona*' (2005) 34 *Journal of Law, Medicine and Ethics*, 398–9.

—— and Nelkin, Dorothy, *Body Bazaar: The Market for Human Tissues in the Biotechnology Age* (New York: Crown, 2001).

Annas, George J., 'Waste and longing: the legal status of placental blood banking' (1999) 340 *New England Journal of Medicine*, 1521–4.

Armitage, S., Warwick, R., Fehily, D., Navarrete, C. and Contreras, M., 'Cord blood banking in London: the first 1000 collections' (1999) 24 *Bone Marrow Transplant*, 139–45.

Bader, Jean–Michel, 'La greffe de visage s'attaque à une grave déformation' ('Face transplants tackle a serious deformity'), *Le Figaro*, 25 January 2007, www.lefigaro.fr/science, accessed 17 April 2007.

Baghan, Alireza, 'Compensated kidney exchange: a review of the Iranian model' (2006) 16 *Kennedy Institute of Ethics Journal*, 269–82.

Barker, Juliet N. and Wagner, John E., 'Umbilical-cord blood transplantation for the treatment of cancer' (2003) 3 *Nature Reviews Cancer*, 526–32.

Barker, J.N., Weisdorf, D.J., DeFor, T.E. et al., 'Rapid and complete donor chimerism in adult recipients of unrelated donor umbilical cord blood transplantation after reduced-intensity conditioning' (2003) 102 *Blood*, 1915–19.

Barnett, Antony, and Smith, Helen, 'Cruel cost of the human egg trade' *Observer*, 30 April 2006, pp. 6–7.

Bartolini, Francesco, Battaglia, Manuela, De Iulio, Cinzia and Sirchia, Girolano, 'Response' (1995) 86 *Blood*, 12, 4900.

Baud, Jean-Pierre, *L'affaire de la main volée: une histoire juridique du corps* (Paris : Éditions du Seuil, 1993).

Baylis, Francoise, 'A face is not just like a hand: *pace* Barker' (2004) 4 *American Journal of Bioethics*, 30–2.

Beeson, Diane, statement before US Congressional Subcommittee on Criminal Justice, Drug Policy and Human Resources, 7 March 2006.

Bell, Daniel, *The End of Ideology: On the Exhaustion of Political Ideas in the Fifties* (New York: The Free Press, 1960).

Benjamin, Walter, 'Capitalism as religion', in Marcus Bullock and Michael W. Jennings (eds), *Selected Writings of Walter Benjamin* (Cambridge, MA: Harvard University Press, 1999), pp. 288–91, originally published 1921.

Berg, Jessica W., 'Risky business: evaluating oocyte donation' (2001) 1 *American Journal of Bioethics*, 4, 18–19.

Beyleveld, Derek and Brownsword, Roger, 'Patenting human genes: legality, morality and human rights', in J.W. Harris (ed.), *Property Problems: From Genes to Pension Funds* (London: Kluwer Law International, 1997), pp. 9–24.

Birke, Lynda, *Feminism and the Biological Body* (Edinburgh: University of Edinburgh Press, 1999).

Blum, Virginia L., *Flesh Wounds: The Culture of Cosmetic Surgery* (Berkeley, CA: University of California Press, 2003).

Boggio, Andrea, 'Charitable trusts and human research genetic databases: the way forward?' (2005) 1 *Genomics, Society and Policy*, 2, 41–9.

Borchadt, John K., 'Children's parents sue over genetics patent' (2000) 1 *The Scientist*, 1122.

Bostyn, S.J.R., 'One patent a day keeps the doctor away? Patenting human genetic information and health care' (2000) 7 *European Journal of Health Law*, 229–64.

Boyle, James, 'Fencing off ideas: enclosure and the disappearance of the public domain', Interactivist Info Exchange, http://slash.autonomedia.org/analysis, accessed 10 September 2004, p. 5.

—— *Shamans, Software and Spleens: Law and the Construction of the Information Society* (Cambridge, MA: Harvard University Press, 1996).

—— 'The second enclosure movement and the construction of the public domain' (2003) 66 *Law and Contemporary Problems*, 33–74.

Brenner, Michael J. et al., 'The spectrum of complications of immunosuppression: is the time right for hand transplantation?' (2002) 84 *Journal of Bone and Joint Surgery*, 1861–70.

British Medical Association, *The Medical Profession and Human Rights: Handbook for a Changing Agenda* (London: Zed Books, 2001), Chapter Eight, 'Trade in Organs', pp. 193–204.

Brody, Baruch, 'Intellectual property and biotechnology—the US internal experience, part 1' (2006) 16 *Kennedy Institute of Ethics Journal*, 1.

Brown, Phyllida, 'Do we even need eggs?' (2006) 439 *Nature*, 655–7.

Brownsword, Roger, 'Biobank governance: business as usual?' (2005) paper presented at the European Commission PropEur workshop, Tuebingen, 20 January.

—— 'Biobank governance: property, privacy and consent', in Christian Lenk, Nils Hoppe and Roberto Andorno (eds), *Ethics and Law of Intellectual Property: Current Problems in Politics, Science and Technology* (Aldershot: Ashgate, 2006), Chapter Five.

Broxmeyer, H.E., 'Biology of cord blood cells and future prospects for enhanced clinical benefit' (2005) 7 *Cytotherapy*, 209–18.

Buckley, Christine, 'Body parts taken from workers at more nuclear sites' (2007) *The Times*, 20 April.

Campbell, Denis, 'Women will be paid to donate eggs for science' (2007) *Observer*, 18 February, p. 1.

Carvel, John, 'With love at Christmas: a set of stem cells' (2005) *Guardian*, 6 December, p. 7.

Caulfield, Timothy, Gold, E. Richard and Cho, Mildred K., 'Patenting human genetic material: refocusing the debate' (2000) 1 *Nature Reviews Genetics*, 227–31.

Cells4Life, 'Response to the Royal College of Obstetricians and Gynaecologists Scientific Advisory Committee Opinion Paper 2 on umbilical cord blood banking', 27 January 2007, www.cells4life.co.uk/news-details, accessed 31 January 2007.

Cheek, Erica, 'Patenting the obvious?' (2007) *Nature*, 3 May, online at www.nature.com, accessed 9 May 2007.

Cheney, Annie, *Body Brokers: Inside America's Underground Trade in Human Remains* (New York: Broadway Books, 2006).

Cherry, Mark, *Kidney for Sale by Owner: Human Organs, Transplantation and the Market* (Georgetown University Press, 2005).

Claeys, Alain, *Rapport sur les conséquences des modes d'appropriation du vivant sur les plans économique, juridique et éthique, Troisiéme partie* (Paris: Office Parlementaire d'Évaluation des Choix Scientifiques et Technologiques, Assemblée Nationale, report no. 1487, www.assemblee-nationale.fre/12/oecst/i11487.asp, accessed 23 September 2004).

Clark, Peter A., 'Face transplantation: Part II—an ethical perspective' (2005) 11 *Medical Science Monitor*, 41–7.

Cobb, Matthew, *The Egg and Sperm Race: The Seventeenth-Century Scientists Who Unravelled the Secrets of Sex, Life and Growth* (London: Free Press, 2006).

Cohen, Alfred, 'Sale or donation of human organs' (2006) *The Journal of Halacha and Contemporary Society*, 37–67.

Cohen, Cynthia B., 'Public policy and the sale of human organs' (2002) 12 *Kennedy Institute of Ethics Journal*, 47–67.

Cohen, G.A., *Self-Ownership, Freedom and Equality* (Cambridge: Cambridge University Press, 1997).

Comité Consultatif National d'Éthique (CCNE), *Opinion No. 1: On sampling of dead human embryonic and foetal tissue for therapeutic, diagnostic and scientific purposes* (22 May 1984), www.ccne-ethique.fr/english/avis/a_001.htm.

—— *Opinion no. 21: That the human body should not be used for commercial purposes* (13 December 1990).

—— *Opinion no. 37: That the human genome should not be used for commercial purposes* (2 December 1991).

—— *Opinion number 74: Umbilical cord blood banks for autologous use or for research* (Paris: CCNE, 2002).

—— *Avis numéro 93: Commercialisation des cellules souches humaines et autres lignées cellulaires* (2006), www.ccne-éthique.fr/français/avis/a_093.htm, accessed 12 April 2007.

Comité Consultatif National d'Éthique (CCNE) and Nationaler Ethikrat (German National Ethics Council), *Opinion No. 77: Ethical problems raised by the collected biological material and associated information data: 'biobanks', 'biolibraries'* (Paris: CCNE, 20 March 2003).

Committee on Human Genome Diversity, Commission on Life Sciences, US National Research Council, *Evaluating Human Genetic Diversity* (National Academy Press, 1997).

Cookson, Clive, 'The cloning connection: cloned tissues from stem cells might beat immune rejection' (2005) www.ScientificAmerican.com, 27 June.

Cooper, Melinda, 'The unborn born again: neo-imperialism, the evangelical right and the culture of life' (2006) 17 *Postmodern Culture*, 1.

Cornish, W.R., Llewelyn, M. and Adcock, M., *Intellectual Property Rights (IPRs) and Genetics: A Study into the Impact and Management of Intellectual Property Rights within the Healthcare Sector* (Cambridge: Cambridge Genetic Knowledge Park, July 2003).

Craven, J. and Rodin, G.M., *Psychiatric Aspects of Organ Transplantation* (New York: Oxford University Press, 1992).

Cyranski, David, 'No end in sight for stem cell odyssey' (2006) 439 *Nature*, 658–9.

Delavigne, A. and Rozenberg, S., 'Epidemiology and prevention of ovarian hyperstimulation syndrome (OHSS): a review' (2002) 8 *Human Reproduction Update*, 559–77.

Derrida, Jacques, *Counterfeit Money* (Chicago: University of Chicago Press, 1992).

Diamond *v.* Chakrabarty, 447 U.S. 303 (1980).

Diamond *v.* Diehr, 450 U.S. 175, 185 (1981).

Dickenson, Donna L., 'Commodification of human tissue: implications for feminist and development ethics' (2002) 2 *Developing World Bioethics*, 1, 55–63.

—— 'Commodification of human tissue: implications for feminist and development ethics', (Portuguese translation) in Debora Diniz (ed.), *Bioética Feminista Contemporânea* (Brazilia: Editora Letras Livres, 2003).

—— 'Patently paradoxical? Public order and genetic patents' (2004) 5 *Nature Reviews Genetics*, 86.

—— 'Procuring gametes for research and therapy: The case for unisex altruism' (1997) 23 *Journal of Medical Ethics*, 93–5.

—— 'Property and women's alienation from their own reproductive labour' (2001) 15 *Bioethics*, 3, 203–17.

—— *Property in the Body: Feminist Perspectives* (Cambridge: Cambridge University Press, 2007).

—— *Property, Women and Politics* (Cambridge: Polity, 1997).

—— *Risk and Luck in Medical Ethics* (Cambridge: Polity, 2003).

—— 'The lady vanishes: what's missing from the stem cell debate' (2006) 3 *Journal of Bioethical Inquiry*, 43–54.

—— 'The threatened trade in human ova' (2004) 5 *Nature Reviews Genetics*, 2, 167.

—— and Vineis, Paolo, 'Evidence-based medicine and quality of care' (2002) 10 *Health Care Analysis*, 3, 243–59.

—— and Widdershoven, Guy, 'Ethical issues in limb transplants' (2001) 15 *Bioethics*, 110–24.

Dienst, Paul van and Savulescu, Julian, 'For and against: no consent should be needed for using leftover body material for scientific purposes' (2002) 325 *British Medical Journal*, 648–51.

Dodds, Susan, 'Women, commodification and embryonic stem cell research', in James Humber and Robert F. Almeder (eds), *Biomedical Ethics Reviews: Stem Cell Research* (Totowa, NJ: Humana Press, 2003), pp. 149–75.

Doodeward *v.* Spence, 6 CLR 406 (1908).

Dubernard, J.M. et al. 'Human hand allograft: report on the first six months' (1999) 353 *Lancet*, 1315–20.

Dunnett, S.B. and Rosser, A.E., 'Cell therapy in Huntington's Disease' (2004) 1 *Neuro Rx*, 394–405.

Ecker, Jeffrey L. and Greene, Michael F., 'The case against private umbilical cord blood banking' (2005) 105 *Obstetrics and Gynecology*, 6, 1282–4.

Edozien, Leroy C., 'NHS maternity units should not encourage commercial banking of umbilical cord blood' (2006) 333 *British Medical Journal*, 801–4.

Eisenberg, Rebecca S., 'How can you patent genes?' (2002) 2 *American Journal of Bioethics*, 3–11.

Elliott, Carl, 'Doing harm, living organ donors, clinical research and *The Tenth Man*' (1995) 21 *Journal of Medical Ethics*, 91–96.

Ende, Norman, 'Letter' (1995) 86 *Blood*, 12, 4699.

English, Veronica, Mussell, Rebecca, Sheather, Julian and Sommerville, Ann, 'Ethics briefings: retention and use of human tissue' (2004) 30 *Journal of Medical Ethics*, 235–6.

Ethics Committee of the American Society for Reproductive Medicine, 'Financial incentives in recruitment of oocyte donors' (2004) 82 *Fertility and Sterility* supplement, S240–4.

European Group on Ethics and New Technologies (EGE), *Opinion on the Ethical Aspects of Umbilical Cord Blood Banking*, opinion number 19, IP/04/364 (Brussels: EGE, 2004).

Evanier, David, 'Parents sue over Canavan test patent' (2001), www.jewishjournal.com, accessed 22 January 2007.

Fabre, Cecile, *Whose Body Is It Anyway? Justice and the Integrity of the Person* (Oxford: Oxford University Press, 2006).

Fagot-Largeault, Anne, 'Ownership of the human body: judicial and legislative responses in France', in Henk ten Have and Jos Welie (eds), *Ownership of the Human Body: Philosophical Considerations on the Use of the Human Body and Its Parts in Healthcare* (Dordrecht: Kluwer, 1998), pp. 115–40.

Fernandez, M.N., Regidor, C. and Cabrera, R., 'Letter: Umbilical cord blood transplantation in adults' (2005) 352 *New England Journal of Medicine*, 935.

FIGO (International Federation of Gynaecology and Obstetrics), *Ethical Guidelines Regarding the Procedure of Collection of Cord Blood* (1998), www.figo.org.

Finegold, David L. et al., *Bioindustry Ethics* (Amsterdam: Elsevier, 2005).

Frank, A.W., 'Emily's scars: surgical shapings' (2004) 34 *Hastings Center Report*, 18–29.

Freeman, Michael and Jaoude, Pauline Abou, 'Justifying surgery's last taboo: the ethics of face transplants' (2007) 33 *Journal of Medical Ethics*, 76–81.

Frow, John, 'Gift and commodity', in John Frow (ed.), *Time and Commodity Culture: Essays in Cultural Theory and Postmodernity* (Oxford: Clarendon, 1997).

Funk Brothers Seed Co. *v.* Kalo Inoculant Co., 333 US 127 (1948).

Gardner, Ralph Jr., 'Looks to die for', *New York* magazine, www.nymag.com, accessed 9 April 2007.

Gilman, Sander L., *Making the Body Beautiful: A Cultural History of Aesthetic Surgery* (Princeton, NJ: Princeton University Press, 1999).

Gitter, Donna M., 'Ownership of human tissue: a proposal for Federal recognition of human research participants' property rights in their biological material' (2004) *Washington and Lee Review*, winter.

Gluckman, E., Broxmeyer, H.A., Auerbach, A.D. et al., 'Hematopoietic reconstitution in a patient with Fanconi's anemia by means of umbilical-cord blood from an HLA-identical sibling' (1989) 321 *New England Journal of Medicine*, 1174–8.

Gold, E. Richard, *Body Parts: Property Rights and the Ownership of Human Biological Materials* (Washington, DC: Georgetown University Press, 1996).

Goldman, Brian, 'HER2: the patent "genee" is out of the bottle' (2007) 176 *Journal of the Canadian Medical Association*, 1443–4.

Goodwin, Michele, *Black Markets: The Supply and Demand of Body Parts* (Cambridge: Cambridge University Press, 2006).

Gorner, Peter, 'Parents suing over patenting of genetic test' (2000) *Chicago Tribune*, 19 November, http://home.iprimus.com.au, accessed 22 January 2007.

Gottlieb, Karen, 'Human biological samples and the law of property: the trust as a model for biological repositories', in R.F. Weir (ed.), *Stored Tissue Samples: Ethical, Legal and Public Policy Implications* (Iowa City: Iowa University Press, 1998), pp. 183–97.

Goyal, M. et al., 'Economic and health consequences of selling a kidney in India' (2003) 288 *Journal of the American Medical Association*, 1589–93.

Greenberg et al. *v.* Miami Children's Hospital Research Institute, Inc., 208 F. Supp. 2d 918 (2002), 264 F Supp 2d 1064 (2003).

Grosjean, Blandine, 'Ces femmes infertiles contraintes à l'exil', *Libération*, 20 August 2004.

Grubb, Andrew, '"I, me mine": bodies, parts and property' (1998) 3 *Medical Law International*, 299–313.

Gunning, Jennifer, 'Umbilical cord blood banking: a surprisingly controversial issue', unpublished report for CCELS (Cardiff Centre for Ethics, Law and Science, n.d.).

Gutmann, Ethan, 'Why Wang Wenyi was shouting: is Beijing committing atrocities against the Falun Gong movement?' (2006) www.weeklystandard.com, 5 August.

Haiken, Elizabeth, *Venus Envy* (Baltimore, MD: Johns Hopkins University Press, 1997).

Haley, Rebecca, Horvath, Liana and Sugarman, Jeremy, 'Ethical issues in cord blood banking: summary of a workshop' (1997) 38 *Tranfusion*, 367–73.

Hall, Stephen S., 'Stem cells: a status report' (2006) 36 *Hastings Center Report*, 16–22.

Hanson, Mark M., 'Religious voices in biotechnology: the case of gene patenting' (1997) 27 *Hastings Center Report*, 1–30.

Hardcastle, Rohan, *Law and the Human Body: Property Rights, Ownership and Control* (Oxford: Hart, 2007).

Hardin, Garrett, 'The tragedy of the commons', 162 *Science*, 1243.

Harris, John, and Erin, C., 'An ethically defensible market in human organs' (2002) 325 *British Medical Journal*, 114–15.

Harrison, Charlotte H., 'Neither Moore nor the market' (2002) 28 *American Journal of Law and Medicine*, 77–104.

Healy, Kieran, *Last Best Gifts: Altruism and the Market for Human Organs* (Chicago: University of Chicago Press, 2006).

Heijnen, Esther, Eijkemans, Marinus, de Klerk, Cora, et al., 'A mild treatment strategy for *in vitro* fertilisation: a randomised non-inferiority trial' (2007) 369 *Lancet*, 743–9.

Heitman, E. and Schlachtenhaufen, M., 'The differential effects of race, ethnicity and socio-economic status on infertility treatment', in C.B. Cohen (ed.), *New Ways of Making Babies: The Case of Egg Donation* (Bloomington, Indiana: Indiana University Press, 1996), pp. 188–212.

Heller, Mark and Eisenberg, Rebecca, 'Can patents deter innovation? The anticommons in biomedical research' (1998) 280 *Science*, 698–701.

Henderson, Mark, 'Cloning benefits oversold, says UK scientist' (2006) *The Times*, 18 December.

Hermitte, Marie-Angèle, *Le sang et le droit: essai sur la transfusion sanguine* (Paris: Éditions du Seuil, 1996).

Highfield, Roger, 'Have we been oversold the stem cell dream?' (2005) *Daily Telegraph*, 29 June.

Hofmeyr, G.K., Bex, P.J.M., Skapinker, R. and Delahunt, T., 'Hasty clamping of the umbilical cord may initiate neonatal intraventricular hemorrhage' (1989) 29 *Medical Hypotheses*, 5.

Holland, Suzanne, 'Contested commodities at both ends of life: buying and selling embryos, gametes and body tissues' (2001) 11 *Kennedy Institute of Ethics Journal*, 283–4.

—— 'Beyond the embryo: a feminist reappraisal of the stem cell debate', in Suzanne Holland and L. Lebacqz (eds), *The Human Embryonic Stem Cell Debate: Science, Ethics and Public Policy* (Cambridge, MA: MIT Press, 2001), pp. 73–86.

Holm, Soren, 'Who should control the use of human embryonic stem cells? A defence of the donor's ability to control' (2006) 3 *Journal of Bioethical Inquiry*, 55–68.

Honigsbaum, Mark, 'Hospitals refuse to warn of bone contamination' (2007) *Guardian*, 6 January, p. 4.

Hoppe, Nils, 'The curse of intangibility: tracing equitable IP rights in human biological material, or *Moore* revisited', in Christian Lenk, Nils Hoppe and Roberto Andorno (eds), *Ethics and Law of Intellectual Property: Current Problems in Politics, Science and Technology* (Aldershot: Ashgate, 2006), Chapter Ten.

Human Fertilisation and Embryology Authority, 'Statement on Donating Eggs for Research', 21 February 2007, www.hfea.gov.uk, accessed 22 February 2007.

Hurlbut, William B., George, Robert P. and Grompe, Markus, 'Seeking consensus: a clarification and defense of altered nuclear transfer' (2006) 36 *Hastings Center Report*, 42–50.

Hutchinson, Roger, *Calum's Road* (Edinburgh: Birlinn, 2006).

Huxtable, Richard and Woodley, Julie, 'Gaining face or losing face? Framing the debate on face transplants' (2005) 19 *Bioethics*, 505–22.

—— '(When) will they have faces? A response to Agich and Siemionov' (2006) 32 *Journal of Medical Ethics*, 403–4.

Hyun, Insoo, 'Fair payment or undue inducement?' (2006) 442 *Nature*, 7103, 629–30.

Hwang, Woo-Suk, et al., 'Evidence of a pluripotent human stem cell line derived from a cloned blastocyst' (2004) 303 *Science*, 1669–74.

—— 'Patient-specific embryonic stem cells derived from human SCNT blastocysts' (2005) 308 *Science*, 1777–83.

Jackson, Emily, 'Fraudulent stem cell research and respect for the embryo' (2006) 1 *Biosocieties*, 349–56.

Jacobs, Allen, Dwyer, James and Lee, Peter, 'Seventy ova' (2001) 31 *Hastings Center Report*, 12–14.

Jenkins, Russell, 'Families demand investigation to end Sellafield's culture of silence' (2007), *The Times*, 18 April.

Jensen, K. and Murray, F., 'International patenting: the landscape of the human genome' (2005) 310 *Science*, 239–40.

Johnson, Josephine, 'Paying egg donors: exploring the arguments' (2006) 36 *Hastings Center Report*, 28–31.

Jones, N.F., 'Concerns about human hand transplantation in the 21st century' (2002) 27 *Journal of Hand Surgery*, 771–87.

Joung, Phillan, 'Breaking the silence: the aftermath of the egg and cloning scandal in South Korea' (2006), paper presented at the conference 'Connecting civil society: implementing basic values', Berlin, 17–19 March.

Juvin, Hervé, *L'avènement du corps* (Paris: Le Débat, Gallimard, 2005).

Kahn, Jeffrey, 'Can we broker eggs without making omelets?' (2001) 1 *American Journal of Bioethics*, 4, 14–15.

Kaiser, Jocelyn, 'Court decides tissue samples belong to university, not patients' (2006) 312 *Science*, 436.

Kang, K.S., Sim, K.W., Yuh, O.H. et al., 'A 37-year-old spinal cord-injured female patient, transplanted of multipotent stem cells from human UC blood, with improved sensory perception and mobility, both functionally and morphologically: a case study' (2005) 7 *Cytotherapy*, 368–73.

Katz, S. and Kravetz, S., 'Facial plastic surgery for people with Down syndrome: research findings and their professional and social implications' (1989) 94 *American Journal of Mental Retardation*, 101–10.

Kim, B.O., Tian, H., Prasongsukarn, K. et al., 'Cell transplantation improves ventricular function after a myocardial infarction: a preclinical study of human unrestricted somatic stem cells in a porcine model' (2005) 112 *Circulation*, 196–204.

Kittredge, Susan Cooke, 'Dad and the bodysnatchers' (2006) *Sunday Times*, 12 March, p. 3.

Kitzinger, Sheila, *The New Experience of Childbirth* (London: Orion, 2004).

Klapheke, Martin M. et al., 'Psychiatric assessment of candidates for hand transplantation' (2000) 20 *Microsurgery*, 453–7.

Klein, Naomi, *No Logo* (London: Picador, 2000).

Knoppers, Bartha M., 'DNA banking: a retrospective perspective', in J. Burley and J. Harris (eds), *A Companion to Genethics* (Oxford: Blackwell, 2002), pp. 379–86.

—— 'Human genetic material: commodity or gift?' in R.F. Weir (ed.), *Stored Tissue Samples* (Iowa City: Iowa University Press, 1998), 226–35.

—— 'Status, sale and patenting of human genetic material: an international survey' (1999) 1 *Nature Reviews Genetics*, 23.

—— et al., 'Ethical issues in international collaborative research on the human genome: the HGP and the HDGP' (1996) 34 *Genomics*, 272–5.

—— Hirtle, M. and Glass, K.C., 'Commercialization of genetic research and public policy' (1999) 286 *Science*, 5448, 2277–8.

Kogler, G. et al., 'A new human somatic stem cell from placental cord blood with intrinsic pluripotent differentiation potential' (2004) 200 *Journal of Experimental Medicine*, 123–5.

Kolata, Giana, 'Who owns your genes?' (2000) *New York Times*, 15 May.

Kuczynski, Alex, *Beauty Junkies: Inside Our $15 Billion Obsession with Cosmetic Surgery* (New York: Doubleday, 2006).

Lainez Villabona, B. et al., 'Early or late umbilical cord clamping? A systematic review of the literature' (2005) 63 *Anales Pediatria*, 1, 14–21.

Lanzendorf, S.E. et al., 'Use of human gametes obtained from anonymous donors for the production of human embryonic stem cell lines' (2001) 76 *Fertility and Society*, 132–7.

Lanzetta, Marco et al., 'The international registry in hand and composite tissue transplantation' (2005) 79 *Transplantation*, 1210–14.

Laqueur, Walter, *Making Sex: Body and Gender from the Greeks to Freud* (Cambridge, MA: Harvard University Press, 1992).

Latteier, Carolyn, *Breasts: The Woman's Perspective on an American Obsession* (New York: Haworth Press, 1998).

Laughlin, M.J., Eapen, M., Rubinstein, P. et al., 'Outcomes after transplantation of cord blood or bone marrow from unrelated donors in adults with leukemia' (2004) 351 *New England Journal of Medicine*, 2265–75.

Laurie, Graeme, '(Intellectual) property? Let's think about staking a claim to our own genetic samples' (Edinburgh: AHRB Research Centre, 2004).

Lauritzen, Paul, 'Stem cells, biotechnology and human rights: implications for a posthuman future' (2005) 35 *Hastings Center Report*, 2, 25–33.

Lazar, Kay, 'Wonder drug for men alleged to cause harm in women' (1999) *Boston Herald*, 22 August.

Leor, J., Guetta, E., Chouraqui, P. et al., 'Human umbilical cord blood cells: a new alternative for myocardial repair?' (2005) 7 *Cytotherapy*, 251–7.

Levitt, Mairi and Weldon, Sue, 'A well-placed trust? Public perceptions of the governance of DNA databases' (2005) 15 *Critical Public Health*, 311–21.

Lilige, Wolfgang, 'The case for adult stem cell research' (2002) *21st Century Science and Technology Magazine*, winter, online at www.sciencetech.com, accessed 7 March 2007.

Locke, John, *The Second Treatise on Civil Government* (1689), in Howard R. Penniman (ed.), *John Locke: On Politics and Education* (New York: D. Van Nostrand, 1947).

McDonald, S.J. and Abbott, J.M., 'Effects of timing of umbilical cord clamping of term infants on maternal and neonatal outcomes (Protocol)', *The Cochrane Database of Systematic Reviews*, issue 1, art. No. CD004074, first published online 20 January 2003.

McDougall, Dan, 'Wives fall prey to kidney trade' (2007) *Observer*, 18 February, p. 43.

McGee, Glenn, 'Gene patents can be ethical' (1999) 7 *Cambridge Quarterly of Healthcare Ethics*, 417–30.

McGivering, Jill, 'China "selling prisoners' organs"' (2006) BBC News online, 19 April.

McHale, Jean, 'Waste, ownership and bodily products' (2000) 8 *Health Care Analysis*, 2, 123–35.

McLeod, Carolyn and Baylis, Francoise, 'Feminists on the inalienability of human embryos' (2006) 24 *Hypatia*, 1–24.

McMillan-Scott, Edward, 'Secret atrocities of Chinese regime' (2006) *Yorkshire Post*, 13 June.

Macklin, Ruth, 'What is wrong with commodification?' in C.R. Cohen (ed.), *New Ways of Making Babies: The Case of Egg Donation* (Bloomington: Indiana University Press, 1996), pp. 106–121.

MacKenzie, Catriona, 'Conceptions of the body and conceptions of autonomy in bioethics', (2004) paper given at the International Association of Bioethics conference, Sydney, November.

Madigan, Nick, 'Inquiry widens after two arrests in cadaver case at UCLA' (2004) *New York Times*, 9 March, p. A21.

Magnus, David and Cho, Mildred K., 'Issues in oocyte donation for stem cell research' (2005) 308 *Science*, 1747–8.

Maio, Giovanni, 'The cultural specificity of research ethics, or why ethical debate in France is different' (2002) 28 *Journal of Medical Ethics*, 147–50.

Marx, Karl, *Capital*, tr. Samuel Moore and Edward Aveling, ed. Frederick Engels (Moscow: Progress, 1954, original edition 1867).

—— *Early Writings*, tr. and ed. T.B. Bottomore (New York: McGraw-Hill, 1963).

—— *Grundrisse: Foundations of the Critique of Political Economy*, tr. with a foreword by Martin Nicolas (New York: Vintage Books, 1973).

Marzano-Parisoli, Maria M., *Penser le corps* (Paris: Presses Universitaires de France, 2002).

Mason, Ken, and Laurie, Graeme, 'Consent or property? Dealing with the body and its parts in the shadow of Bristol and Alder Hey' (2001) 9 *Medical Law Review*, 710–29.

Matas, David and Kilgour, David, *Bloody Harvest: Revised Report into Allegations of Organ Harvesting from Falun Gong Prisoners in China*, 31 January 2007, p. 27, downloaded from www.investigation.go.saveinter.net, accessed 14 May 2007.

Mauss, Marcel, *The Gift: The Form and Reason for Exchange in Archaic Societies* (London: Routledge, 1990).

Mead, H.M., *Tikanga Maori: Living by Maori Values* (Wellington, New Zealand. Huia Publishers, 2003).

Medical Research Council, *Human Tissue and Biological Samples for Use in Research: Operational and Ethical Guidelines* (London: MRC, 2001).

—— *Public Perceptions of the Collection of Human Biological Samples* (London: MRC, 2000).

Memmi, Dominique, *Les gardiens du corps: dix ans de magistère bioéthique* (Paris: Éditions de l'École des Hautes Études en Sciences Sociales, 1996).

Merleau-Ponty, Maurice, *The Visible and the Invisible* (Evanston, IL: Northwestern University Press, 1968).

Moise, Kenneth J. Jr., 'Umbilical cord stem cells' (2005) 106 *Obstetrics and Gynecology*, 1393–407.

Momberger, K., 'Breeder at law' (2002) 11 *Columbia Journal of Gender and Law*, 127–74.

Moore *v.* Regents of the University of California, 51 Cal 3rd 120, 793 P 2d, 271 Cal Rptr 146 (1990). Cert. Denied 111 SCt 1388.

Morley, G.M., 'Cord closure: can hasty clamping injure the newborn?' (1998) *Obstetrics and Gynaecology Management*, July.

Munzer, Stephen R., *A Theory of Property* (Cambridge: Cambridge University Press, 1990).

—— 'An uneasy case against property rights in human body parts' (1994) 11 *Social Philosophy and Policy*, 2, 259–86.

—— 'The special case of property rights in umbilical cord blood for transplantation' (1999) 51 *Rutgers Law Review*, 493–568.

Nelkin, Dorothy, 'Is bioethics for sale?' (2003) 24 *The Tocqueville Review*, 2, 45–60.

Nelkin, Dorothy, and Lindee, Susan, *The DNA Mystique: The Gene as a Cultural Icon* (New York: W.H. Freeman and Co., 1995).

Nicholls, Heidi, 'Cord blood used to make miniature artificial liver' (2006) *Bionews Today*, 11 November, www.bionews.org.uk, accessed 30 January 2007.

Normile, Dennis, and Mann, Charles, 'Asia jockeys for stem cell lead' (2005) 307 *Science*, 660–4.

Norsigian, J., 'Egg donation for IVF and stem cell research: Time to weigh the risks to women's health' (2005) in Boston Women's Health Collective (ed.), *Our Bodies Ourselves*, ch. 25, online at www.ourbodiesourselves./orgbook/companion.

Nuffield Council on Bioethics, *Stem Cell Therapies: The Ethical Issues* (London: Nuffield Council on Bioethics, 2000).

—— *The Ethics of Patenting DNA* (London: Nuffield Council on Bioethics, 2002).

Oldham, Paul, 'The patenting of plant and animal genomes' (2006) paper presented at workshop of the European Commission project PropEur, Paris, May.

Ossorio, Pilar, 'Legal and ethical issues in biotechnology patenting', in J. Burley and J. Harris (eds), *A Companion to Genethics* (Oxford: Blackwell, 2002), pp. 408–19.

Otten, J., Wyle, H. and Phelps, G., 'The charitable trust as a model for genomic banks' (2004) 350 *New England Journal of Medicine*, 85–6.

Paik, Young-Gyung, 'Beyond bioethics: the globalized reality of ova trafficking and the possibility of feminist intervention' (2006) paper presented at the International Forum on the Human Rights of Women and Biotechnology, Seoul, 21 September.

Paradise, Jordan, Andrews, Lori and Holbrook, Timothy, 'Patents on human genes: an analysis of scope and claims' (2006) 307 *Science*, 1566–7.

Parfit, Tom, 'Beauty salons fuel trade in aborted babies' (2005) *Guardian Unlimited*, 17 April, online at www.guardian.co.uk.

Parry, Bronwyn, *Trading the Genome: Investigating the Commodification of Bio-Information* (New York: Columbia University Press, 2004).

Pearson, Helen, 'Health effects of egg donation may take decades to emerge' (2006) 442 *Nature*, 7103, 607–8.

Pennings, G. and DeVroey, M., 'Subsidized IVF treatment and the effect on the number of egg sharers' (2006) 1 *Reproductive Medicine Online*, 8–10.

Plato, *The Republic*, tr. Paul Shorey, in *The Collected Dialogues of Plato including the Letters*, Edith Hamilton and Huntington Cairns (eds) (New York: Pantheon Books, 1961).

Pollock, Anne, 'Complicating power in high-tech reproduction' (2003) 24 *Journal of Medical Humanities*, 241–63.

Prendiville, W. and Elbourne, D., 'Care during the third stage of labour', in

I. Chalmers, M. Enkin and M.J.N.C. Keirse (eds), *Effective Care in Pregnancy and Childbirth* (Oxford: Oxford University Press, 1989), pp. 1145–69.

Prendiville, W.J., Elbourne, D. and McDonald, S., 'Active versus expectant management in the third stage of labour', *The Cochrane Database of Systematic Reviews* (2000) issue 3, Art. No. CD000007 (24 July).

Proctor, S.J., Dickinson, A.M., Parekh, T. and Chapman, C., 'Umbilical cord blood banks in the UK have proved their worth and now deserve a firmer foundation' (2001) 323 *British Medical Journal*, 60–1.

R. *v.* Kelly (1998) 3 All ER 741.

Rabe, H., Reynolds, G. and Diaz-Rossello, J., 'Early versus delayed umbilical cord clamping in pre-term infants', *The Cochrane Database of Systematic Reviews* (2004) issue 4, art. No. CD003248pub2, first published 18 October 2004, with more recent review in volume 3, 17 May 2005.

Rabinow, Paul, *French DNA: Trouble in Purgatory* (Chicago: University of Chicago Press, 1999).

Radcliffe Richards, Janet, 'The case for allowing kidney sales' (1998) 352 *Lancet*, 1950–2.

Radin, Margaret J., *Contested Commodities: The Trouble with Trade in Sex, Children, Body Parts and Other Things* (Cambridge, MA: Harvard University Press, 1996).

Rai, Nirmala and Prathima Reddy, letter, 'Rapid Response' to Leroy C. Edozien, 'NHS maternity units should not encourage commercial banking of umbilical cord blood' (2006) 333 *British Medical Journal*, 801–4, www.bmj.com/cgi/letters, accessed 31 January 2007.

Randerson, James, 'Rise and fall of clone king who doctored stem-cell research' (2005) *Guardian*, 24 December.

—— 'IVF women will be paid to offer spare eggs for research' (2006) *Guardian*, 28 July, p. 12.

Resnik, David, 'Regulating the market for human eggs' (2001) 15 *Bioethics*, 1, 1–26.

—— 'The commercialization of human stem cells: ethical and policy issues' (2002) 10 *Health Care Analysis*, 127–54.

Retained Organs Commission, *A Consultation Document on Unclaimed and Unidentifiable Organs and Tissue: A Possible Regulatory Framework* (NHS, February 2002).

Rocha, Vanderson, et al., 'Graft-versus-host disease in children who have received a cord-blood or bone marrow transplant from an HLA-identical sibling' (2000) 342 *New England Journal of Medicine*, 25, 1846–54.

—— Labopin, M., Sans, G. et al., 'Transplants of umbilical cord blood or bone marrow from unrelated donors in adults with leukemia' (2004) 351 *New England Journal of Medicine*, 2276–85.

Rogers, Ian, and Casper, Robert F., 'Lifeline in an ethical quagmire: umbilical cord blood as an alternative to embryonic stem cells' (2004) 2 *Sexuality, Reproduction and Menopause*, 2, 64–70.

Rose, Carol M., 'The comedy of the commons: customs, commerce and inherently public property' (1986) 53 *University of Chicago Law Review*, 742.

Rose, Hilary, 'An ethical dilemma: the rise and fall of UmanGenomics—the model biotech company?' (2004) 425 *Nature*, 123–4.

Rothstein, Mark A., 'Expanding the ethical analysis of biobanks' (2005) 33 *Journal of Law, Medicine and Ethics*, 1, 89–101.

Royal College of Obstetricians and Gynaecologists Scientific Advisory Committee, *Opinion Paper 2: Cord Blood Banking* (London: RCOG, 2001).

—— *Umbilical Cord Blood Banking*, Scientific Advisory Committee Opinion Paper 2 (London: RCOG, 2006).

Royal College of Surgeons Working Party, *Facial Transplantation* (London: Royal College of Surgeons, 2006, second edition).

Rubinstein, P., Rosenfeld, R.E., Adamson, J.W. and Stevens, C.E., 'Stored placental blood for unrelated bone marrow reconstitution' (1993) 81 *Blood*, 1679–90.

Sample, Ian, 'Stem cell pioneer accused of faking all his research. Apart from the cloned dog' (2006) *Guardian*, 11 January, p. 11.

—— 'Branson launches shared stem cell bank' (2007) *Guardian*, 2 February, p. 11.

Sanner, M.A., 'People's feelings and ideas about receiving transplants of different origins: questions of life and death, identity and nature's border' (2001) 15 *Clinical Transplant*, 19–27.

Sauer, M.V., 'Defining the incidence of serious complications experienced by oocyte donors: a review of 1,000 cases' (2001) 184 *American Journal of Obstetrics and Gynecology*, 277–8.

—— 'Egg donor solicitation: problems exist, but do abuses?' (2001) 1 *American Journal of Bioethics*, 4, 1–2.

—— 'Indecent proposal: $5,000 is not "reasonable compensation" for oocyte donors [editorial]' (1999) 71 *Fertility and Sterility*, 7–8.

Sauer, M.V., Paulson, R.J. and Lobo, R.A., 'Rare occurrence of ovarian hyperstimulation syndrome in oocyte donors' (1996) 52 *International Journal of Gynecology and Obstetrics*, 259–62.

Savulescu, Julian, 'Is the sale of body parts wrong?' (2003) 16 *Journal of Medical Ethics*, 117–19.

Scheper-Hughes, Nancy, 'Bodies for sale: whole or in parts' (2002) 7 *Body and Society*, 1–8.

Schneider, Ingrid and Schumann, Claudia, 'Stem cells, therapeutic cloning, embryo research: women as raw material suppliers for science and industry', in Svea Luise Herrmann and Margaretha Kurmann (eds), *Reproductive Medicine and Genetic Engineering: Women between Self-Determination and Societal Standardisation*, proceedings of a conference held in Berlin, 15-17 November 2001 (Reprokult, 2002), pp. 70–9.

Schneider, Susan Weidman, 'Jewish women's eggs: a hot commodity in the IVF marketplace' (2001) 26 *Lilith*, 3, 22.

Senituli, Lopeti, 'They came for sandalwood, now the b ... s are after our genes!', paper presented at the conference 'Research ethics, Tikanga Maori/indigenous and protocols for working with communities' (2004) Wellington, New Zealand, 10–12 June.

Sève, Lucien, *Qu'est-ce que la personne humaine? Bioéthique et démocratie* (Paris: La Dispute, 2006).

Sexton, Sarah, 'Transforming "waste" into "resource": from women's eggs to economics for women' (2005), paper presented at Reprokult workshop, Heinrich Boll Foundation, Berlin, 10 September.

Sharp, Lesley A., *Bodies, Commodities and Biotechnologies: Death, Mourning and Scientific Desire in the Realm of Human Organ Transfer* (New York: Columbia University Press, 2007).

Shaw, M. (ed.), *After Barney Clark: Reflections on the Utah Artificial Heart Program* (Austin, TX: University of Texas Press, 1984).

Shiffrin, Seana Valentine, 'Lockean arguments for private intellectual property', in Stephen R. Munzer (ed.), *New Essays in the Legal and Political Theory of Property* (Cambridge: Cambridge University Press, 2002), pp. 138–67.

Shiva, Vandana, *Biopiracy: The Plunder of Nature and Knowledge* (Boston: South End Press, 1997).

Sidaway *v.* Bethlem RHG [1985] 1 All ER 643.

Skene, Loane, 'Ownership of human tissue and the law' (2002) 3 *Nature Reviews Genetics*, 145–8.

Skloot, Rebecca, 'Big news on the tissue research front: a congressional investigation into researchers profiting off tissues without consent' (2006) 'Culture Dish' weblog, 14 June, http://rebeccaskloot.blogspot.com, accessed 27 March 2007.

—— 'Henrietta's Dance' (2000) *Johns Hopkins Magazine*, April, www.jhu.edu/~jhumag/0400web/01.html, accessed 27 March 2007.

—— 'Taking the least of you: the tissue-industrial complex' (2006) *New York*

Times, 16 April, available at www.nytimes.com/2006/04/16/magazine/16tissuehtml, accessed 24 April 2006.

—— 'Tissue ownership update, II: more on *Catalona*' (2006) www.rebeccaskloot.blog.spot.com, 18 April, accessed 27 March 2007.

Smyth, Anna, 'A leap in the dark?' (2007) *Scotsman*, 2 February, www.news.scotsman.com, accessed 5 February 2007.

Solves, P., Moraga, R., Saucedo, E. et al., 'Comparison between two strategies for umbilical cord blood collection' (2003) 31 *Bone Marrow Transplant*, 269–73.

Souder, Mark, 'Human cloning and embryonic stem cell research after Seoul: examining exploitation, fraud and ethical problems in the research' (2006) Opening Statement of Chairman, Congressional Subcommittee on Criminal Justice, Drug Policy and Human Resources, 7 March.

Spar, Debora, *The Baby Business: How Money, Science and Politics Drive the Commerce of Conception* (Cambridge, MA: Harvard Business School Press, 2006).

Steigenga, M.J. et al., 'Evolutionary conserved structures as indicators of medical risk: increased incidence of cervical ribs after ovarian hyperstimulation in mice' (2006) 56 *Animal Biology*, 1, 63–8.

Steinbock, Bonnie, 'Payment for egg donation and surrogacy' (2004) 71 *Mount Sinai Journal of Medicine*, 255–65.

Steinbrook, Robert, 'Egg donation and human embryonic stem-cell research' (2006) 354 *New England Journal of Medicine*, 324–6.

Stercx, Sigrid and Cockbain, Julian, 'Stem cell patents and morality: the European Patent Office's emerging policy', forthcoming.

Stock, Gregory, 'Eggs for sale: How much is too much?' (2001) 1 *American Journal of Bioethics*, 4, 26–7.

Sugarman, Jeremy, Reisner, Emily G. and Kurtzberg, Joanne, 'Ethical issues of banking placental blood for transplantation' (1995) 274 *Journal of the American Medical Association*, 1763–85.

Tallis, Raymond, *The Hand: A Philosophical Inquiry into Human Being* (Edinburgh: Edinburgh University Press, 2003).

Titmuss, Richard M., *The Gift Relationship: From Human Blood to Social Policy*, Ann Oakley and J. Ashton (eds) (London: LSE Books, 1997, 2nd edition).

Thompson, Charis, 'Why we should, in fact, pay for egg donation' (2007) 2 *Reproductive Medicine*, 203–9.

Tromp, Saskia, *Seize the Day, Seize the Cord*, unpublished undergraduate medical dissertation, University of Maastricht (2001).

United Kingdom Department of Health, *Human Bodies, Human Choices:*

The Law on Human Organs and Tissue in England and Wales, a Consultation Report (London: DOH, 2002).

van Diest, P.J. and Savulescu, Julian, 'For and against: no consent should be needed for using leftover body material for scientific purposes' (2002) 325 *British Medical Journal*, 648–51.

van Dusen, Susan, 'Whose genes are they?', online at Chicago Jewish Community/Jewish United Fund website, www.juf.org.news, accessed 22 January 2007.

van Rheenen, Patrick and Bernard J. Brabin, 'Late umbilical cord-clamping as an intervention for reducing iron deficiency anaemia in term infants in developing and industrialised countries: a systematic review' (2004) 24 *Annals of Tropical Paediatrics*, 3–16.

Vermot Mangold, Ruth-Gaby, *Trafficking in Organs in Europe*, report no. 9822 (Strasbourg: Council of Europe, 2003).

Vogel, Gretchen, 'Ethical oocytes: available for a price' (2006) 313 *Science*, 5784, 155.

Waldby, Catherine and Mitchell, Robert, *Tissue Economies: Blood, Organs and Cell Lines in Late Capitalism* (Durham, NC: Duke University Press, 2006).

Wallace, Susan and Stewart, Alison, 'Cord blood banking: guidelines and prospects' (2004) Cambridge Genetic Knowledge Park report, 22 November, online at www.cambridgenetwork.co.uk/pooled/articles, accessed 19 May 2005.

Washington University *v.* William J. Catalona, 437 F Supp 2d, ESCD Ed Mo 2006.

Wiemels, J.L., Cazzaniga, G., Daniotti, M., Eden, O.B., Addison, G.M., Masera, G. et al., 'Prenatal origin of acute lymphoblastic leukaemia in children' (1999) 352 *Lancet*, 1499–503.

Wiggins, O.P., Barker, J.H., Martinez, S. et al., 'On the ethics of facial transplantation' (2004) 4 *American Journal of Bioethics*, 1–12.

Wilkinson, Stephen, *Bodies for Sale: Ethics and Exploitation in the Human Body Trade* (London: Routledge, 2003).

—— 'Commodification arguments for the legal prohibition of organ sale' (2000) 8 *Health Care Analysis*, 189–201.

Williams, Rachel, 'Four accused of stealing Cooke's bones' (2006) *Guardian*, 24 February, p. 1.

Winickoff, David E. and Neumann, Larissa B., 'Towards a social contract for genomics: property and the public in the "biotrust" model' (2005) 1 *Genomics, Society and Policy*, 3, 8–21.

—— and Winickoff, Richard N., 'The charitable trust as a model for genomic biobanks' (2003) 349 *New England Journal of Medicine*, 12, 1180–4.

Yelling, A. (ed.), *Common Field and Enclosure in England, 1450–1850* (Hamden, CT: Archon Books, 1977).

Yoshino, Kimi, 'Fertility scandal at UC Irvine is far from over'(2006) *Los Angeles Times*, 22 January.

Younge, Gary, 'Embryo scientist quits team over ethics fear' (2003) *Guardian*, 14 November, p. 19.

—— and Jones, Sam, 'Family's dismay after Alastair Cooke's bones stolen by New York gang'(2005) *Guardian*, 23 December, p. 3.

Zilberstein, Moshe, Feingold, Michael and Selbel, Machelle M., 'Umbilical cord-blood banking: lessons learned from gamete donation' (1997) 349 *Lancet*, 642–5.

Zimmerman, D. et al., 'Gender disparity in living renal transplant donation' (2000) 36 *American Journal of Kidney Diseases*, 534–40.

Index